SpringerBriefs in Molecular Science

SpringerBriefs in Molecular Science present concise summaries of cutting-edge research and practical applications across a wide spectrum of fields centered around chemistry. Featuring compact volumes of 50 to 125 pages, the series covers a range of content from professional to academic. Typical topics might include:

- A timely report of state-of-the-art analytical techniques
- A bridge between new research results, as published in journal articles, and a contextual literature review
- A snapshot of a hot or emerging topic
- An in-depth case study
- A presentation of core concepts that students must understand in order to make independent contributions

Briefs allow authors to present their ideas and readers to absorb them with minimal time investment. Briefs will be published as part of Springer's eBook collection, with millions of users worldwide. In addition, Briefs will be available for individual print and electronic purchase. Briefs are characterized by fast, global electronic dissemination, standard publishing contracts, easy-to-use manuscript preparation and formatting guidelines, and expedited production schedules. Both solicited and unsolicited manuscripts are considered for publication in this series.

Tatiana Q Aguiar · Eduardo Coelho ·
Lucília Domingues

Ashbya gossypii in Biotechnology

 Springer

Tatiana Q Aguiar
CEB—Centre of Biological Engineering
University of Minho
Braga, Portugal

LABBELS—Associate Laboratory
Braga/Guimarães, Portugal

Lucília Domingues
CEB—Centre of Biological Engineering
University of Minho
Braga, Portugal

LABBELS—Associate Laboratory
Braga/Guimarães, Portugal

Eduardo Coelho
CEB—Centre of Biological Engineering
University of Minho
Braga, Portugal

LABBELS—Associate Laboratory
Braga/Guimarães, Portugal

ISSN 2191-5407 ISSN 2191-5415 (electronic)
SpringerBriefs in Molecular Science
ISBN 978-3-032-12434-0 ISBN 978-3-032-12435-7 (eBook)
https://doi.org/10.1007/978-3-032-12435-7

This Springer imprint is published by the registered company Springer Nature Switzerland AG
The registered company address is: Gewerbestrasse 11, 6330 Cham, Switzerland

If disposing of this product, please recycle the paper.

Preface

The story of *Ashbya gossypii* in science began in 1926. Mine began much later, in the early 2000s, just after I had finished my Ph.D., and involved two renowned and truly inspirational scientists, Merja Penttilä from VTT, Finland, and Peter Phillipsen from Biozentrum, University of Basel, Switzerland. I still recall a dinner with Merja, my former supervisor at VTT and a researcher with remarkable experience in Biotechnology. When I asked if she could think of a research topic worth pursuing, she paused and mentioned a fungus with unusual features, among them the presence of lipid droplets that had long fascinated a colleague of hers, Peter, whom I had the pleasure of meeting some years later. The name of that fungus was *Ashbya gossypii*. That conversation sparked my curiosity, and as soon as I searched for *Ashbya*, I realized its potential to go far beyond its established role in riboflavin production. That was the beginning of my own journey with this extraordinary organism.

Over the past two decades, I have been fortunate to share this path with brilliant Ph.D. students, Orquídea Ribeiro, Tatiana Q Aguiar, Rui Silva (co-supervised by Tatiana Q Aguiar), and more recently Miguel Francisco (co-supervised by Eduardo Coelho), and of receiving financial support from the Fundação para a Ciência e Tecnologia (FCT), through projects such as "AshByofactory—*Ashbya gossypii*: a systems metabolic engineered cell factory" and "EssenTial—Establishing sustainable bioproduction of lactones from metabolic engineering of industrial cell factory systems: *Ashbya gossypii*." The present book is, in many ways, a natural outcome of this journey. I would like to warmly thank my two co-authors, Eduardo Coelho and Tatiana Q Aguiar, for embracing this challenge with me and for sharing, and spreading, the enthusiasm for *Ashbya* over the years. My gratitude also goes to Carlos Costa, who not only contributed as a researcher within the EssenTial project but is also the creative mind behind the figures in this book. I am further indebted to the many researchers who, in addition to those already mentioned, have collaborated on our projects, as well as to the broader *Ashbya* scientific community, whose collective efforts have laid the foundation for what this book has become.

This volume, *Ashbya gossypii in Biotechnology*, provides the first comprehensive account of *A. gossypii* as a biotechnological production platform. The book brings together the molecular, metabolic, and process engineering strategies that have transformed this filamentous fungus from a niche model organism into a promising host for industrial biotechnology. Its scope is wide-ranging: from the history of *Ashbya* (Chap. 1), through physiology and metabolism (Chap. 2), to advances in metabolic engineering (Chap. 3), bioprocessing strategies (Chap. 4), and finally, future perspectives (Chap. 5). Together, these chapters paint a complete picture of how *Ashbya* is being developed as a sustainable microbial platform for green chemistry and bio-based production.

The book is intended for researchers, biotechnologists, and students in microbiology and metabolic engineering who seek both a foundational reference and a roadmap for innovation. It arrives at a particularly important moment. As the world faces urgent challenges in sustainability and the circular economy, new microbial cell factories are needed to deliver environmentally friendly solutions. In this context, *A. gossypii* stands out as a strong player, with its proven ability in riboflavin overproduction and its emerging potential for a wider range of biotechnologically relevant products.

Looking ahead, I am confident that *A. gossypii* will continue to affirm its role in industrial biotechnology. As the world increasingly seeks bio-based solutions within a circular economy, *Ashbya* stands out as a versatile and robust microbial platform. It is my hope that this book will not only serve as a comprehensive reference for researchers and students but also inspire new ideas and applications that will shape the future of sustainable bioproduction.

Braga, Portugal   Lucília Domingues

Acknowledgments This study was supported by the Portuguese Foundation for Science and Technology (FCT) under the scope of the strategic funding of UID/04469/2025 unit (DOI: 10.54499/UID/04469/2025), the project ESSEntial (DOI: 10.54499/PTDC/BII-BTI/1858/2021), the TQ Aguiar assistant research contract (DOI: 10.54499/2023.07897.CEECIND/CP2841/CT0031), and by LABBELS—Associate Laboratory in Biotechnology, Bioengineering and Microelectromechanical Systems, LA/P/0029/2020. Authors also acknowledge the project VIIAFOOD for its Assistant Research program under the scope of "Agenda para a Inovação Empresarial—ViiaFood—Plataforma de Valorização, Industrialização e Inovação comercial para o Agro-alimentar" (Project no. 37, application no. C644929456-00000040), funded by the Plan for Recovery and Resilience (PRR) and by the European Funds Next Generation EU.

The authors also acknowledge the LABBELS' Science Communication Officer Carlos Costa for generating the images in this book and the use of ChatGPT (https://chatgpt.com/) for language editing and writing style improvement purposes.

Competing Interests The authors have no competing interests to declare that are relevant to the content of this manuscript.

Contents

Chapter 1
Ashbya gossypii History

Abstract This chapter traces the evolution of *Ashbya gossypii* from a cotton pathogen into a versatile microbial platform for industrial biotechnology. Initially identified in the 1920s as the causative agent of stigmatomycosis, *A. gossypii* was later shown to rely on insect vectors for plant colonization, revealing a symbiotic lifestyle. Its natural capacity to overproduce riboflavin (vitamin B2) attracted early industrial interest and laid the foundation for decades of applied research. Despite early taxonomic uncertainties, molecular and genomic studies uncovered a close evolutionary relationship with *Saccharomyces cerevisiae*. This, combined with a compact genome and efficient homologous recombination, established *A. gossypii* as a genetically tractable model organism. Advances in transformation and gene-targeting tools during the 1990s enabled precise metabolic engineering, cementing its leading role in sustainable industrial riboflavin production. From the 2010s onward, continued advances in metabolic engineering and synthetic biology broadened its industrial applications beyond riboflavin production. Since then, *A. gossypii* has been engineered to produce a growing portfolio of valuable bioproducts, including nucleosides, orotic acid, folates, biolipids, lactones, terpenes, gangliosides, and recombinant proteins. This remarkable journey from plant pathogen to industrial microbial cell factory highlights the central role of *A. gossypii* in advancing sustainable "white biotechnology".

Keywords *Ashbya gossypii* · Habitat · Ecology · Taxonomy · Model organism · Microbial cell factory · Riboflavin production · Industrial biotechnology

A. gossypii first gained scientific awareness in **1926**, when Ashby and Nowell isolated it from infected cotton bolls (*Gossypium* spp.) in the British West Indies, identifying it as a pathogen responsible for stigmatomycosis—a destructive disease of cotton plants (Ashby and Nowell 1926). In the subsequent decades, it was found infecting seeds and young fruits of other important tropical and subtropical crops, such as beans, citrus, coffee, and tomatoes (Batra 1973; Pridham and Raper 1950). Controlled inoculation studies confirmed that *A. gossypii* cannot penetrate intact plant tissues on its own, relying on piercing-sucking insects, particularly those of the sub-order Heteroptera, to

T. Q. Aguiar et al., *Ashbya gossypii in Biotechnology*,
SpringerBriefs in Molecular Science, https://doi.org/10.1007/978-3-032-12435-7_1

introduce its spores or mycelium into plant wounds (Batra 1973; Pridham and Raper 1950). Given this dependency on insect vectors, targeted insecticide use effectively reduced *A. gossypii* dissemination and stigmatomycosis incidence (Dammer and Ravelo 1990). In **2013**, Dietrich et al. (2013) reported the isolation of *A. gossypii* strains from milkweed bugs feeding on oleander in Florida (USA), an important finding that demonstrated its sustained ecological persistence and further illuminated its tight insect-associated lifestyle.

Ecologically, *A. gossypii* does not need to degrade complex substrates, thriving in plant tissues rich in simple sugars and oils (Caskey and Gallup 1931; Pridham and Raper 1950; Stahmann et al. 1994), thereby its modest extracellular enzyme secretion (Aguiar et al. 2014). Instead, the fungus produces strong volatile aromas that attract insect vectors, which facilitate its dispersal (Ravasio et al. 2014; Wendland and Walther 2011; Silva et al. 2019a; Birk et al. 2019; Semenova et al. 2022). Moreover, its natural propensity for riboflavin overproduction, first reported in **1946** (Wickerhan et al. 1946), is thought to benefit both *A. gossypii* and insect vectors, by neutralizing reactive oxygen species generated during early plant immune responses to pathogen infection (Walther and Wendland 2012; Dietrich et al. 2013; Silva et al. 2019b).

Taxonomically, *A. gossypii* was first described as *Nematospora gossypii* in **1926**, owing to its resemblance to other *Nematospora* species (Ashby and Nowell 1926). Two years later, Guilliermond reassigned it to a newly created genus—*Ashbya*—within the class Hemiascomycetes, based on its filamentous growth and multinucleate cells (Guilliermond 1928). Simultaneously, Ciferri and Fragoso (1928) also proposed reassigning it to a new genus, *Ashbia*, placing it under the order Saccharomycetales. However, since the name *Ashbya* had been published one month earlier, it was given priority over the Latinized version *Ashbia*. Along the years, the distinctive needle-shaped ascospores and exclusive hyphal growth were features that clearly set *Ashbya* apart from typical Saccharomycetales yeasts (von Arx and van der Walt 1987). However, early genomic insights in the **mid-1990s** revealed surprisingly high gene homology and gene order conservation between *A. gossypii* and *Saccharomyces cerevisiae*, pointing to a much closer evolutionary relationship than previously thought (Steiner and Philippsen 1994; Altmann-Jöhl and Philippsen 1996). At the same time, based on ribossomal DNA (rDNA)-based phylogenetic analyses, Kurtzman (Kurtzman 1995) proposed merging *Ashbya, Eremothecium, Holleya,* and *Nematospora* into a single genus, *Eremothecium,* under the new family Eremotheciaceae. The studies of Messner et al. (Messner et al. 1995) and Prillinger et al. (Prillinger et al. 1997) reinforced this taxonomic grouping, but aligned the genus with the Saccharomycetaceae family. A most definitive classification milestone came in **2003–2004**, when the Philippsen group published the genome sequence of *A. gossypii*, which revealed a compact 9 Mb genome organized in seven chromosomes encoding approximately 4776 protein-coding genes, 90% of which with syntenic homologs in *S. cerevisiae* (Brachat et al. 2003; Dietrich et al. 2004; 2013). The absence of transposons, the presence of only seven chromosomes, and the high gene order conservation suggested the species diverged before the whole-genome duplication event seen in *S. cerevisiae* (Dietrich et al. 2004, 2013; Wendland and Walther 2011). Subsequent phylogenomic studies confirmed the positioning of *A.*

gossypii among pre-duplication yeast species like *Kluyveromyces lactis, K. waltii* and *S. kluyveri*, though debates on the monophyly of these four species persist (Kurtzman and Robnett 2003; Fitzpatrick et al. 2006; Wang et al. 2009). Despite the Kurtzman taxonomic assignment of *A. gossypii* to the *Eremothecium* genus in 1995, the scientific community continued on using preferably the homotypic synonym defined by Guilliermond (1928), *Ashbya gossypii* (Kurtzman and de Hoog 2011). In **2013**, a new *Ashbya* species was isolated from insects feeding on trees of the genus *Acer*, which was named *A. aceri* (Dietrich et al. 2013). Since *A. gossypii* and *A. aceri* display much closer resemblances between each other than with other *Eremothecium* species (Altmann-Jöhl and Philippsen 1996; Wendland and Walther 2011; Perez-Nadales et al. 2014), it might well be that future taxonomic reassignments consider placing these species again under the independent *Ashbya* genus proposed by Guilliermond (1928). Indeed, based on pairwise sequence similarities, Malimas et al. (2023) recently reassigned five *Eremothecium* species to four resurrected genera: *Ashbya, Crebrothecium, Holleya,* and *Nematospora*. However, since all recognized *Eremothecium* species form a monophyletic group and share a common ecological trait, Li et al. (2025) subsequently argued that it remains both practical and scientifically justifiable to retain them in the genus *Eremothecium*, as proposed by Kurtzman (1995). Nevertheless, their multilocus phylogenetic analysis of 65 new isolated strains revealed two novel species, *E. koelreuteriae* and *E. leptocoridis*, which form a clade more closely related to *A. gossypii* and *A. aceri* than to other *Eremothecium* species (Li et al. 2025), further highlighting the taxonomic heterogeneity within this genus. Noteworthy, several reports have also demonstrated strain heterogeneity within the *A. gossypii* species itself (Ribeiro et al. 2012; Dietrich et al. 2013; Ledesma-Amaro et al. 2015; Silva et al. 2019a).

Alongside its ecological and taxonomic journey, *A. gossypii*'s physiology attracted attention. Early studies in the **1930s** documented the production of a yellow pigment in aging cultures of certain strains (Farries and Bell 1930), later identified as riboflavin (vitamin B2) (Wickerhan et al. 1946). During the **1940s** and early **1950s** further physiological studies proliferated in the literature as a reflection of the eventual interest in using this fungus for industrial riboflavin production, as a more genetically stable alternative to the initially explored *Eremothecium ashbyi* (Tanner et al. 1949; Pfeifer et al. 1950; Mickelson 1950; Smiley et al. 1951; Pridham and Raper 1950; Goodman and Ferrera 1954; Demain 1972). Starting in the **1950s**, industrial processes leveraged low-cost nutrient sources like corn steep liquor, peptones, and vegetable oils (Tanner et al. 1949; Pridham and Raper 1950; Pfeifer et al. 1950; Mickelson 1950; Smiley et al. 1951; Malzahn et al. 1959), while later decades employed mutagenesis and fermentation optimization (Pridham and Raper 1952; Demain 1972; Perlman 1979; Lago and Kaplan 1981; Bigelis 1989; Schmidt et al. 1996; Park et al. 2007; Tajima et al. 2009; Dessipri et al. 2022). From the **late 1950s**, Grain Processing Corp pioneered a biotechnological process for riboflavin production using *A. gossypii*, which was discontinued around 1968 due to lack of competitiveness with the chemical synthesis process (Malzahn et al. 1959; Lago and Kaplan, 1981). It was not until **1974** that a competitive riboflavin production process employing *A. gossypii* was established by Merck (Lago and Kaplan 1981), which continued working on strain improvement by classical mutagenesis, generating strains able to yield higher than 15 g/L riboflavin (Bigelis 1981; Demain 2007). In **1990**, large-scale riboflavin production using *A. gossypii* became fully industrialized when BASF, after acquiring the bioprocess developed by Merck, launched its commercial production in Germany (Stahman et al. 2000). After operating both chemical and biotechnological riboflavin production processes in parallel for six years, BASF discontinued their long-term running chemical process due to the superior economic and environmental sustainability of the biotechnological process (Stahman et al. 2000; Wall-Markowski et al. 2004; Wenda et al. 2011), thereby presenting a successful example of the sustainable "white biotechnology" enabled by *A. gossypii*. In **2003**, production was relocated to a new site in Gunsan, South Korea (Schwechheimer et al. 2016). Over the years BASF continued to develop industrially competitive *A. gossypii* strains, supported by advances in modern metabolic engineering approaches (Aguiar et al. 2015; Revuelta et al. 2017; Dessipri et al. 2022).

A molecular toolbox for the targeted and rational engineering of *A. gossypii* began taking shape in the **early 1990s**, spearheaded by the Philippsen group with the development of transformation methods (Wright and Philippsen 1991; Altmann-Jöhl and Philippsen 1996). A first breakthrough was the discovery that plasmids containing *S. cerevisiae* ARS elements (*ARS1* and 2-micron) could freely replicate in *A. gossypii* (Wright and Philippsen 1991), enabling gene overexpression. A subsequent major breakthrough was the discovery of the efficient homologous recombination system in *A. gossypii* (Steiner et al. 1995), which enabled targeted gene disruption (Altmann-Jöhl and Philippsen 1996) and gene integration (Althöfer et al. 1999). This advancement laid the foundation for targeted metabolic engineering in this organism, pioneered by the Revuelta group through the overexpression of genes involved in riboflavin biosynthesis (*AgRIBs*) (Althöfer et al. 1999; Althöfer and Revuelta 2003). The identification of strong constitutive promoters, such as those from the *A. gossypii* translation elongation factor 1α (*AgTEF*) and glyceraldehyde-3-phosphate dehydrogenase (*AgGPD*) genes (Steiner and Philippsen 1994; Revuelta et al. 1999), further supported robust expression of homologous and heterologous genes. In **2000**, another major advance came from the Philippsen group with the development of simple PCR-based gene-targeting techniques for the genetic manipulation of *A. gossypii* (Wendland et al. 2000). This was supported by the development of several heterologous dominant selectable markers, greatly expanding the *A. gossypii* genetic toolkit and enabling more flexible and efficient engineering strategies (reviewed in Aguiar et al. 2015). Together, these advances established *A. gossypii* as a genetically tractable and industrially relevant host, with BASF using only self-cloned *A. gossypii* strains in large-scale production since 2001 (Revuelta et al. 2017; Dessipri et al. 2022). Its high genetic tractability, characterized by efficient homologous recombination and gene-targeting, combined with its close relationship to yeast-like fungi and small genome size, promoted the adoption of *A. gossypii* as a model system for studying fungal growth, development and genome evolution from the **mid-2000s** onward (Wendland and Walther 2005; Schmitz and Philippsen 2011; Perez-Nadales et al. 2014). Supported by continued expansion of the *A. gossypii* molecular toolbox throughout the **2010s**, these foundational developments marked a significant milestone in the emergence of *A. gossypii* as a powerful microbial platform for industrial biotechnology (Aguiar et al. 2015).

Following a decade of significant contributions by the Stahmann, Revuelta and Park groups toward improving *A. gossypii* strains for riboflavin production (Aguiar et al. 2015; Revuelta et al. 2017), the **2010s** saw a paradigm shift in *A. gossypii* biotechnology toward the production of other economically relevant bioproducts, driven by the efforts of the Revuelta/Jiménez and Domingues groups (Aguiar et al. 2015, 2017). Among these feature recombinant proteins (Ribeiro et al. 2010, 2013; Magalhães et al. 2014; Aguiar et al. 2014), biolipids (Ledesma-Amaro et al. 2014a, b, 2018; Lozano-Martínez et al. 2017; Díaz-Fernández et al. 2017, 2019), nucleosides (Ledesma-Amaro et al. 2015; 2016), folates (Serrano-Amatriain et al. 2016) and lactones (Silva et al. 2019a). During the **2020s** newly explored bioproducts have further expanded this list: flavin adenine dinucleotide (FAD) (Patel and Chandra 2020), orotic acid (Silva et al. 2022), monoterpenes (Muñoz-Fernández et al. 2022, 2024) and gangliosides (Montero-Bullón et al. 2025). Today, *A. gossypii* stands as a successful example of sustainable "white biotechnology", a once-plant pathogen transformed into a cornerstone of industrial vitamin B2 production, with ongoing developments in systems biology and metabolic engineering paving the way for new biotechnological applications (Aguiar et al. 2017; Ledesma-Amaro et al. 2018; Silva et al. 2019a, 2022; Díaz-Fernández et al. 2019; Patel and Chandra 2020; Muñoz-Fernández et al. 2022; Montero-Bullón et al. 2025). Throughout this journey (Fig. 1.1), numerous researchers have contributed to the development of the *A. gossypii* biotechnology. The most frequently involved contributors are acknowledged in the list below, ordered chronologically by their first publication involving this fungus.

Ashby SF (1926)	Park EY (2001)
Nowell W (1926)	Walther A (2003)
Tanner FW (1948)	Jiménez A (2005)
Pridham TG (1950)	Kato T (2006)
Raper KB (1950)	Sugimoto T (2009)
Demain AL (1970)	Chandra TS (2009)
Özbas T (1986)	Domingues L (2010)
Kutsal T (1986)	Penttilä M (2010)
Philippsen P (1991)	Wiebe MG (2010)
Steiner S (1994)	Aguiar TQ (2013)
Sahm H (1994)	Ledesma-Amaro R (2013)
Stahmann KP (1994)	Silva R (2015)
Wendland J (1995)	Buey RM (2015)
Schmidt G (1996)	Wittmann C (2015)
Revuelta JL (1999)	Muñoz-Fernández G (2020)
Santos MA (1999)	Montero-Bullón JF (2021)
Dietrich FS (1999)	

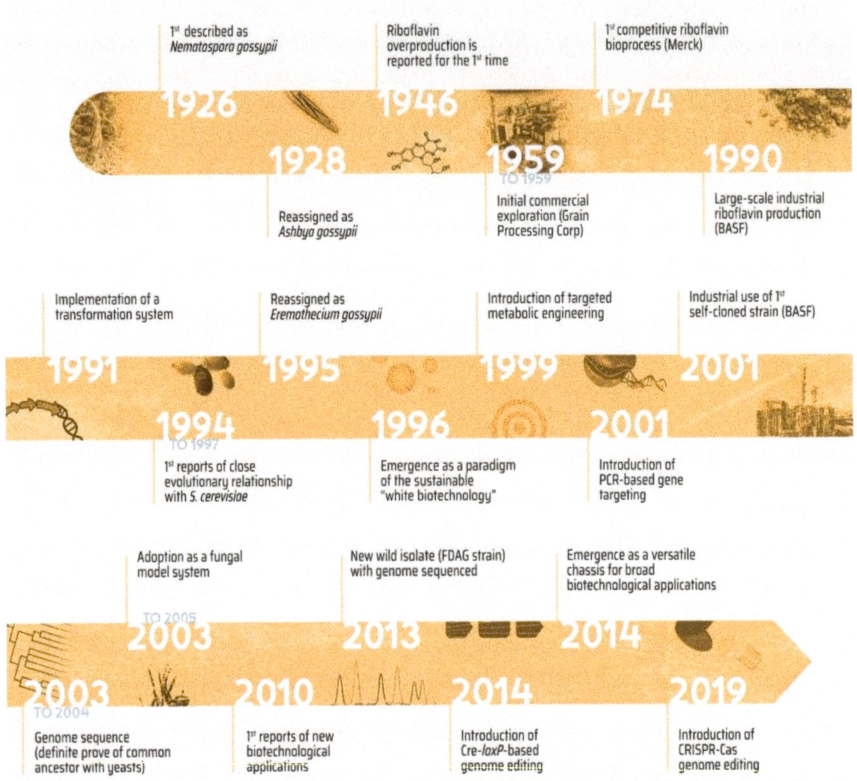

Fig. 1.1 Timeline highlighting key milestones in the development of *Ashbya gossypii* as a biotechnological platform

References

Aguiar TQ, Ribeiro O, Arvas M, Wiebe MG, Penttilä M, Domingues L (2014) Investigation of protein secretion and secretion stress in *Ashbya gossypii*. BMC Genomics 15:1137. https://doi.org/10.1186/1471-2164-15-1137

Aguiar TQ, Silva R, Domingues L (2015) *Ashbya gossypii* beyond industrial riboflavin production: a historical perspective and emerging biotechnological applications. Biotechnol Adv 33:1774–1786. https://doi.org/10.1016/j.biotechadv.2015.10.001

Aguiar TQ, Silva R, Domingues L (2017) New biotechnological applications for *Ashbya gossypii*: challenges and perspectives. Bioengineered 8:309–315. https://doi.org/10.1080/21655979.2016.1234543

Althöfer H, Revuelta JL (2003) Genetic strain optimization for improving the production of riboflavin. WO Patent 2003048367 A1, 12 Jun 2003. https://patents.google.com/patent/WO2003048367A1/en

Althöfer H, Seulberg H, Zelder O, Revuelta JL (1999) Genetic method for producing riboflavin. WO Patent 1999061623 A2, 2 Dec 1999. https://patents.google.com/patent/WO1999061623A2/en

Altmann-Jöhl R, Philippsen P (1996) *AgTHR4*, a new selection marker for transformation of the filamentous fungus *Ashbya gossypii*, maps in a four-gene cluster that is conserved between *A. gossypii* and *Saccharomyces cerevisiae*. Mol Gen Genet 250:69–80. https://doi.org/10.1007/BF02191826

Ashby SF, Nowell W (1926) The fungi of stigmatomycosis. Ann Bot 40:69–84. https://doi.org/10.1093/oxfordjournals.aob.a090018

Batra LR (1973) Nematosporaceae (Hemiascomycetidae): Taxonomy, pathogenicity, distribution, and vector relations. USDA Technical Bulletin 1469, Washington DC, p 71. https://dn790003.ca.archive.org/0/items/bub_gb_Q6coAAAAYAAJ_2/bub_gb_Q6coAAAAYAAJ.pdf

Bigelis R (1989) Industrial products of biotechnology: Application of gene technology. In: Jacobson GK, Jolly SO (eds) Biotechnology: Gene Technology, vol 7B. VCH, Weinheim, pp 230–259

Birk F, Fraatz MA, Esch P, Heiles S, Pelzer R, Zorn H (2019) Industrial riboflavin fermentation broths represent a diverse source of natural saturated and unsaturated lactones. J Agric Food Chem 67:13460–13469. https://doi.org/10.1021/acs.jafc.9b01154

Brachat S, Dietrich FS, Voegeli S, Zhang Z, Stuart L, Lerch A, Gates K, Gaffney T, Philippsen P (2003) Reinvestigation of the *Saccharomyces cerevisiae* genome annotation by comparison to the genome of a related fungus: *Ashbya gossypii*. Genome Biol 4:R45. https://doi.org/10.1186/gb-2003-4-7-r45

Caskey C, Gallup WD (1931) Changes in the sugar, oil and gossypol content of the developing cotton boll. J Agric Res 42:671–673. https://ia801509.us.archive.org/34/items/in.ernet.dli.2015.33738/2015.33738.Journal-Of-Agricultural-Research-Vol-42-1931.pdf

Ciferri R, Fragoso RG (1928) Hongos parásitos y saprofitos de la República Dominicana (16.ª serie). Bol Real Soc Esp Hist Nat 28:377–388. https://bibdigital.rjb.csic.es/viewer/10600/?offset=#page=386%26viewer=picture%26o=bookmark%26n=0%26q=

Dammer K-H, Ravelo HG (1990) Verseuchung von *Leptoglossus gonagra* (Fabr.) mit *Nematospora coryli* Peglion und *Ashbya gossypii* (Ashby et Nowell) Guilliermond in einer Zitrusanlage der Republik Kuba. Arch Phytopathol Pflanzenschutz 26:71–78. https://doi.org/10.1080/03235409009438945

Demain AL (1972) Riboflavin oversynthesis. Annu Rev Microbiol 26:369–388. https://doi.org/10.1146/annurev.mi.26.100172.002101

Demain AL (2007) The business of biotechnology. Ind. Biotechnol 3:269–283. https://doi.org/10.1089/ind.2007.3.269

Dessipri E, Smith J, Walch S, Srinivasan J (2022) Riboflavin from *Ashbya gossypii*. In: 92nd JECFA—Chemical and Technical Assessment (CTA). FAO, Rome, p 13. https://openknowledge.fao.org/server/api/core/bitstreams/70910278-fc3b-460e-9197-a131d5477619/content

Díaz-Fernández D, Lozano-Martínez P, Buey RM, Revuelta JL, Jiménez A (2017) Utilization of xylose by engineered strains of *Ashbya gossypii* for the production of microbial oils. Biotechnol Biofuels 10:3. https://doi.org/10.1186/s13068-016-0685-9

Díaz-Fernández D, Aguiar TQ, Martín VI, Romaní A, Silva R, Domingues L, Revuelta JL, Jiménez A (2019) Microbial lipids from industrial wastes using xylose-utilizing *Ashbya gossypii* strains. Bioresour Technol 293:122054. https://doi.org/10.1016/j.biortech.2019.122054

Dietrich FS, Voegeli S, Brachat S, Lerch A, Gates K, Steiner S, Mohr C, Pöhlmann R, Luedi P, Choi S, Wing RA, Flavier A, Gaffney TD, Philippsen P (2004) The *Ashbya gossypii* genome as a tool for mapping the ancient *Saccharomyces cerevisiae* genome. Science 304:304–307. https://doi.org/10.1126/science.1095781

Dietrich FS, Voegelli S, Kuo S, Philippsen P (2013) Genomes of *Ashbya* fungi isolated from insects reveal four mating-type loci, numerous translocations, lack of transposons, and distinct gene duplications. G3 (Bethesda) 3:1225–1239. https://doi.org/10.1534/g3.112.002881

Farries EHM, Bell AF (1930) On the metabolism of *Nematospora gossypii* and related fungi, with special reference to the source of nitrogen. Ann Bot 44:423–455. https://doi.org/10.1093/oxfordjournals.aob.a090228

Fitzpatrick DA, Logue ME, Stajich JE, Butler G (2006) A fungal phylogeny based on 42 complete genomes derived from supertree and combined gene analysis. BMC Evol Biol 6:99. https://doi.org/10.1186/1471-2148-6-99

Goodman JJ, Ferrera RR (1954) Synthesis of riboflavin by *Ashbya gossypii* grown in a synthetic medium. Mycologia 46:556–563. https://doi.org/10.1080/00275514.1954.12024396

Guilliermond MA (1928) Recherches sur quelques Ascomycètes inférieurs isolés de la Stigmatomycose des graines de cotonnier. Essai sur la phylogénie des Ascomycètes. Rev Gen Bot 40:328–342, 397–414, 474–485, 555–574, 606–624, 690–704. https://gallica.bnf.fr/ark:/12148/bpt6k54537421/f379.item

Kurtzman CP (1995) Relationships among the genera *Ashbya*, *Eremothecium*, *Holleya*, and *Nematospora* determined from rDNA sequence divergence. J Industr Microbiol 14:523–530. https://doi.org/10.1007/BF01573968

Kurtzman CP, Robnett CJ (2003) Phylogenetic relationships among yeasts of the 'Saccharomyces complex' determined from multigene sequence analyses. FEMS Yeast Res 3:417–432. https://doi.org/10.1016/s1567-1356(03)00012-6

Kurtzman CP, de Hoog GS (2011) *Eremothecium* Borzi emend. Kurtzman (1995). In: Kurtzman CP, Fell JW, Boekhout T (eds) The yeasts: a taxonomic study, 5th edn. Elsevier, San Diego, pp 405–412. https://doi.org/10.1016/B978-0-444-52149-1.00030-6

Lago BD, Kaplan L (1981) Vitamin fermentations: B2 and B12. In: Moo-Young M, Vezina C, Singh K (eds) Advances in biotechnology: fermentation products, vol 3. Pergamon Press, Oxford, pp 241–246. https://doi.org/10.1016/B978-0-08-025385-5.50045-6

Ledesma-Amaro R, Santos MA, Jiménez A, Revuelta JL (2014a) Strain design of *Ashbya gossypii* for single-cell oil production. Appl Environ Microbiol 80:1237–1244. https://doi.org/10.1128/aem.03560-13

Ledesma-Amaro R, Santos MA, Jiménez A, Revuelta JL (2014b) Tuning single-cell oil production in *Ashbya gossypii* by engineering the elongation and desaturation systems. Biotechnol Bioeng 111:1782–1791. https://doi.org/10.1002/bit.25245

Ledesma-Amaro R, Buey RM, Revuelta JL (2015) Increased production of inosine and guanosine by means of metabolic engineering of the purine pathway in *Ashbya gossypii*. Microb Cell Fact 14:58. https://doi.org/10.1186/s12934-015-0234-4

Ledesma-Amaro R, Buey RM, Revuelta JL (2016) The filamentous fungus *Ashbya gossypii* as a competitive industrial inosine producer. Biotechnol Bioeng 113:2060–2063. https://doi.org/10.1002/bit.25965

Ledesma-Amaro R, Jiménez A, Revuelta JL (2018) Pathway grafting for polyunsaturated fatty acids production in *Ashbya gossypii* through golden gate rapid assembly. ACS Synth Biol 7:2340–2347. https://doi.org/10.1021/acssynbio.8b00287

Li CH, Ou JH, Chen CY (2025) *Eremothecium* species from *Cardiospermum halicacabum* and *Koelreuteria henryi*, and their associated soapberry bugs. Mycol Progress 24:55. https://doi.org/10.1007/s11557-025-02073-4

Lozano-Martínez P, Buey RM, Ledesma-Amaro R, Jiménez A, Revuelta JL (2017) Engineering *Ashbya gossypii* strains for *de novo* lipid production using industrial by-products. Microb Biotechnol 10:425–433. https://doi.org/10.1111/1751-7915.12487

Magalhães F, Aguiar TQ, Oliveira C, Domingues L (2014) High-level expression of *Aspergillus niger* β-galactosidase in *Ashbya gossypii*. Biotechnol Prog 30:261–268. https://doi.org/10.1002/btpr.1844

Malimas T, Vu HTL, Yukphan P, Tanasupawat S, Yamada Y (2023) The subdivision of the genus *Eremothecium* Borzi emend. Kurtzman (1995). Int J Novel Res Life Sci 10:58–63. https://doi.org/10.5281/zenodo.10041023

Malzahn RC, Phillips RF, Hanson AM (1959) Riboflavin process. US Patent 2,876,169, 3 May 1959. https://patents.google.com/patent/US2876169A/en

Messner R, Prillinger H, Ibl M, Himmler G (1995) Sequences of ribosomal genes and internal transcribed spacers move three plant parasitic fungi, *Eremothecium ashbyi*, *Ashbya gossypii*, and *Nematospora coryli*, towards *Saccharomyces cerevisiae*. J Gen Appl Microbiol 41:31–42. https://doi.org/10.2323/jgam.41.31

Mickelson MN (1950) The metabolism of glucose by *Ashbya gossypii*. J Bacteriol 59:559–566. https://doi.org/10.1128/jb.59.5.659-666.1950

Montero-Bullón JF, Martín-González J, Ledesma-Amaro R, Jiménez A, Buey RM (2025) Pioneering microbial synthesis of gangliosides in the filamentous fungus *Ashbya gossypii*. Biotechnol Biofuels Bioprod 18:95. https://doi.org/10.1186/s13068-025-02697-4

Muñoz-Fernández G, Martínez-Buey R, Revuelta JL, Jiménez A (2022) Metabolic engineering of *Ashbya gossypii* for limonene production from xylose. Biotechnol Biofuels Bioprod 15:79. https://doi.org/10.1186/s13068-022-02176-0

Muñoz-Fernández G, Montero-Bullón JF, Martínez JL, Buey RM, Jiménez A (2024) *Ashbya gossypii* as a versatile platform to produce sabinene from agro-industrial wastes. Fungal Biol Biotechnol 11:16. https://doi.org/10.1186/s40694-024-00186-1

Park EY, Zhang JH, Tajima S, Dwiarti L (2007) Isolation of *Ashbya gossypii* mutant for an improved riboflavin production targeting for biorefinery technology. J Appl Microbiol 103:468–476. https://doi.org/10.1111/j.1365-2672.2006.03264.x

Patel MV, Chandra TS (2020) Metabolic engineering of *Ashbya gossypii* for enhanced FAD production through promoter replacement of *FMN1* gene. Enzyme Microb Technol 133:109455. https://doi.org/10.1016/j.enzmictec.2019.109455

Perez-Nadales E, Nogueira MFA, Baldin C, Castanheira S, El Ghalid M, Grund E, Lengeler K, Marchegiani E, Mehrotra PV, Moretti M, Naik V, Oses-Ruiz M, Oskarsson T, Schäfer K, Wasserstrom L, Brakhage AA, Gow NA, Kahmann R, Lebrun MH, Perez-Martin J, Di Pietro A, Talbot NJ, Toquin V, Walther A, Wendland J (2014) Fungal model systems and the elucidation of pathogenicity determinants. Fungal Genet Biol 70:42–67. https://doi.org/10.1016/j.fgb.2014.06.011

Perlman D (1979) Microbial process for riboflavin production. In: Pepplev HJ, Pevlman D (eds) Microbial technology, vol 1. Academic, New York, pp 521–527. https://doi.org/10.1016/B978-0-12-551501-6.50021-8

Pfeifer VF, Tanner FW, Vojnovich C, Traufler DH (1950) Riboflavin by fermentation with *Ashbya gossypii*. Ind Eng Chem 42:1776–1781. https://doi.org/10.1021/ie50489a027

Pridham TG, Raper KB (1950) *Ashbya gossypii* – Its significance in nature and in the laboratory. Mycologia 42:603–623. https://doi.org/10.1080/00275514.1950.12017863

Pridham TG, Raper KB (1952) Studies on variation and mutation in *Ashbya gossypii*. Mycologia 44:452–469. https://doi.org/10.1080/00275514.1952.12024209

Prillinger H, Schweigkofler W, Breitenbach M, Briza P, Staudacher E, Lopandic K, Molnár O, Weigang F, Ibl M, Ellinger A (1997) Phytopathogenic filamentous (*Ashbya, Eremothecium*) and dimorphic fungi (*Holleya, Nematospora*) with needle-shaped ascospores as new members within the Saccharomycetaceae. Yeast 13:945–960. https://doi.org/10.1002/(SICI)1097-0061(199708)13:10%3c945::AID-YEA150%3e3.0.CO;2-5

Ravasio D, Wendland J, Walther A (2014) Major contribution of the Ehrlich pathway for 2-phenylethanol/rose flavor production in *Ashbya gossypii*. FEMS Yeast Res 14:833–844. https://doi.org/10.1111/1567-1364.12172

Revuelta JL, Ledesma-Amaro R, Lozano-Martinez P, Díaz-Fernández D, Buey RM, Jiménez A (2017) Bioproduction of riboflavin: a bright yellow history. J Ind Microbiol Biotechnol 44:659–665. https://doi.org/10.1007/s10295-016-1842-7

Revuelta JL, Santos MA, Pompejus M, Seulberger H (1999) Promoter from *Ashbya gossypii*. WO Patent 1999033993 A1, 8 Jul 1999. https://patents.google.com/patent/WO1999033993A1/en

Ribeiro O, Wiebe M, Ilmén M, Domingues L, Penttilä M (2010) Expression of *Trichoderma reesei* cellulases CBHI and EGI in *Ashbya gossypii*. Appl Microbiol Biotechnol 87:1437–1446. https://doi.org/10.1007/s00253-010-2610-7

Ribeiro O, Domingues L, Penttilä M, Wiebe M (2012) Nutritional requirements and strain heterogeneity in *Ashbya gossypii*. J Basic Microbiol 52:582–589. https://doi.org/10.1002/jobm.201100383

Ribeiro O, Magalhães F, Aguiar TQ, Wiebe MG, Penttilä M, Domingues L (2013) Random and direct mutagenesis to enhance protein secretion in *Ashbya gossypii*. Bioengineered 4:1–10. https://doi.org/10.4161/bioe.24653

Schmidt G, Stahmann KP, Kaesler B, Sahm H (1996) Correlation of isocitrate lyase activity and riboflavin formation in the riboflavin overproducer *Ashbya gossypii*. Microbiology 142:419–426. https://doi.org/10.1099/13500872-142-2-419

Schmitz HP, Philippsen P (2011) Evolution of multinucleated *Ashbya gossypii* hyphae from a budding yeast-like ancestor. Fungal Biol 115:557–568. https://doi.org/10.1016/j.funbio.2011.02.015

Schwechheimer SK, Park EY, Revuelta JL, Becker J, Wittmann C (2016) Biotechnology of riboflavin. Appl Microbiol Biotechnol 100:2107–2119. https://doi.org/10.1007/s00253-015-7256-z

Semenova E, Presniakova V, Kozlovskaya V, Markelova N, Gusev A, Linert W, Kurakov A, Shpichka A (2022) The *in vitro* cytotoxicity of *Eremothecium* oil and its components—aromatic and acyclic monoterpene alcohols. Int J Mol Sci 23:3364. https://doi.org/10.3390/ijms23063364

Serrano-Amatriain C, Ledesma-Amaro R, López-Nicolás R, Ros G, Jiménez A, Revuelta JL (2016) Folic acid production by engineered *Ashbya gossypii*. Metab Eng 38:473–482. https://doi.org/10.1016/j.ymben.2016.10.011

Silva R, Aguiar TQ, Coelho E, Jiménez A, Revuelta JL, Domingues L (2019a) Metabolic engineering of *Ashbya gossypii* for deciphering the *de novo* biosynthesis of γ-lactones. Microb Cell Fact 18:62. https://doi.org/10.1186/s12934-019-1113-1

Silva R, Aguiar TQ, Domingues L (2022) Orotic acid production from crude glycerol by engineered *Ashbya gossypii*. Bioresour Technol Rep 17:100992. https://doi.org/10.1016/j.biteb.2022.100992

Silva R, Aguiar TQ, Oliveira R, Domingues L (2019b) Light exposure during growth increases riboflavin production, reactive oxygen species accumulation and DNA damage in *Ashbya gossypii* riboflavin-overproducing strains. FEMS Yeast Res 19:foy114. https://doi.org/10.1093/femsyr/foy114

Smiley KL, Sobolov M, Austin FL, Rasmussen RA, Smith MB, van Lanen JM, Stone L, Boruff CS (1951) Biosynthesis of riboflavin, *L. bulgaricus* factor, and other growth factors. Laboratory and pilot plant studies of biosynthesis by *A. gossypii* cultivated on grain stillage media. Ind Eng Chem 43:1380–1384. https://doi.org/10.1021/ie50498a033

Stahmann KP, Kupp C, Feldmann SD, Sahm H (1994) Formation and degradation of lipid bodies found in the riboflavin-producing fungus *Ashbya gossypii*. Appl Microbiol Biotechnol 42:121–127. https://doi.org/10.1007/BF00170234

Stahmann KP, Revuelta JL, Suelberger H (2000) Three biotechnical process using *Ashbya gossypii*, *Candida famata*, or *Bacillus subtilis* compete with chemical riboflavin production. Appl Microbiol Biotechnol 53:509–516. https://doi.org/10.1007/s002530051649

Steiner S, Philippsen P (1994) Sequence and promoter analysis of the highly expressed *TEF* gene of the filamentous fungus *Ashbya gossypii*. Mol Gen Genet 242:263–271. https://doi.org/10.1007/BF00280415

Steiner S, Wendland J, Wright MC, Philippsen P (1995) Homologous recombination as the main mechanism for DNA integration and cause of rearrangements in the filamentous ascomycete *Ashbya gossypii*. Genetics 140:973–987. https://doi.org/10.1093/genetics/140.3.973

Tajima S, Itoh Y, Sugimoto T, Kato T, Park EY (2009) Increased riboflavin production from activated bleaching earth by a mutant strain of *Ashbya gossypii*. J Biosci Bioeng 108:325–329. https://doi.org/10.1016/j.jbiosc.2009.04.021

Tanner FW, Vojnovich C, Van Lanen JM (1949) Factors affecting riboflavin production by *Ashbya gossypii*. J Bacteriol 58:737–745. https://doi.org/10.1128/jb.58.6.737-745.1949

von Arx JA, van der Walt JP (1987) Ophiostomatales and Endomycetales. Stud Mycol 30:167–176. https://www.studiesinmycology.org/sim/Sim30/fulltext/index.html?i=13

Wall-Markowski CA, Kicherer A, Saling P (2004) Using eco-efficiency analysis to assess renewable-resource–based technologies. Environ Prog 23:329–333. https://doi.org/10.1002/ep.10051

Walther A, Wendland J (2012) Yap1-dependent oxidative stress response provides a link to riboflavin production in *Ashbya gossypii*. Fungal Genet Biol 49:697–707. https://doi.org/10.1016/j.fgb.2012.06.006

Wang H, Xu Z, Gao L, Hao B (2009) A fungal phylogeny based on 82 complete genomes using the composition vector method. BMC Evol Biol 9:195. https://doi.org/10.1186/1471-2148-9-195

Wenda S, Illner S, Mell A, Kragl U (2011) Industrial biotechnology—the future of green chemistry? Green Chem 13:3007–3047. https://doi.org/10.1039/C1GC15579B

Wendland J, Walther A (2005) *Ashbya gossypii*: a model for fungal developmental biology. Nat Rev Microbiol 3:421–429. https://doi.org/10.1038/nrmicro1148

Wendland J, Ayad-Durieux Y, Knechtle P, Rebischung C, Philippsen P (2000) PCR-based gene targeting in the filamentous fungus *Ashbya gossypii*. Gene 242:381–391. https://doi.org/10.1016/s0378-1119(99)00509-0

Wendland J, Walther A (2011) Genome evolution in the *Eremothecium* clade of the *Saccharomyces* complex revealed by comparative genomics. G3 (Bethesda) 1:539–548. https://doi.org/10.1534/g3.111.001032

Wickerham LS, Flickinger MH, Johnson RM (1946) Production of riboflavin by *Ashbya gossypii*. Arch Biochem 9:95–98

Wright MC, Philippsen P (1991) Replicative transformation of the filamentous fungus *Ashbya gossypii* with plasmids containing *Saccharomyces cerevisiae* ARS elements. Gene 109:99–105. https://doi.org/10.1016/0378-1119(91)90593-z

Chapter 2
Ashbya gossypii Physiology and Metabolism

Abstract This chapter provides a comprehensive review on *Ashbya gossypii* physiology and metabolism, with a particular emphasis on the characteristic traits of this filamentous fungus and its natural metabolic capabilities. It is divided into three subsections. The first subsection covers growth and reproduction, describing the mechanisms of spore germination, hyphal growth, septation and sporulation to perform a full coverage of the fungus lifecycle. The second subsection dives into the natural metabolism of *Ashbya gossypii*, covering central carbon and nitrogen metabolism with a particular focus on its genetic and metabolic distinctive traits. The third subsection explores in greater detail the natural metabolic capabilities of *A. gossypii* zooming in on different metabolites of biotechnological interest, by performing a detailed review of the main natural metabolites and the biosynthetic pathways involved in their production. With this critical review, this chapter sets the knowledge base that serves as foundation for understanding the potential of this organism for biotechnological development.

Keywords *Ashbya gossypii* · Lifecycle · Physiology · Native metabolism · Biosynthetic pathways · Riboflavin · Nucleosides · Folates · Microbial lipids · Lactones · Terpenes · Higher alcohols

2.1 Growth and Reproduction

As most filamentous fungi, *A. gossypii* propagates via sporulation, with the spore being the starting point for the formation of a new mycelium (Fig. 2.1). Spore germination initiates with isotropic growth and the formation of a germ bubble (Wendland 2020), which shifts to polar growth for the emergence of a lateral germ tube, growing perpendicularly to the axis of the spore needle (Knechtle et al. 2003). This is followed by the formation of a second germ tube on the opposite side of the germling in an event of bipolar branching, from which hyphae will form and propagate in both directions (Schmitz and Philippsen 2011). During the early stages of growth, hyphae will start to exhibit novel axes of polarity, leading to the perpendicular lateral branching

T. Q. Aguiar et al., *Ashbya gossypii in Biotechnology*,
SpringerBriefs in Molecular Science, https://doi.org/10.1007/978-3-032-12435-7_2

of new hyphae which will themselves extend via polar growth. Hyphal expansion, measured as the tip speed during polar growth, increases throughout growth leading to the dichotomous branching of hyphae into Y-shaped structures, creating new polar growing tips and further expansion of the mycelium (Ayad-Durieux et al. 2000).

As in other filamentous fungi, endogenous protein secretion in *A. gossypii* primarily occurs at actively growing hyphal tips. While hyphal growth is typically associated with enhanced protein secretion, *A. gossypii*'s capacity to secrete both endogenous and heterologous proteins to the extracellular space is more comparable to that of yeast than to classical filamentous fungi (Ribeiro et al. 2010, 2013; Aguiar et al. 2014a, b; Magalhães et al. 2014). *A. gossypii* also produces yeast-like *N*-glycan structures intermediate in length between those of *S. cerevisiae* and other filamentous fungal species (Aguiar et al. 2013). Still, despite its close phylogenetic relationship with *S. cerevisiae*, the regulation of the protein secretion and glycosylation pathways

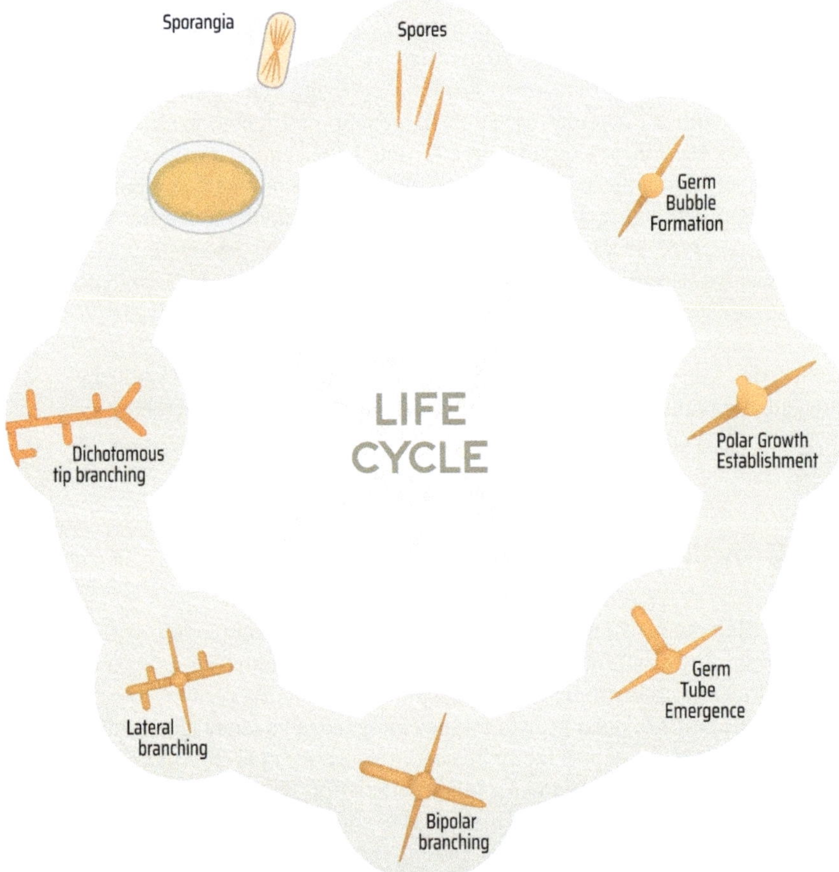

Fig. 2.1 Representation of the *Ashbya gossypii* lifecycle

in *A. gossypii* differs significantly, likely as reflection of their critical roles for hyphal growth (Aguiar et al. 2013, 2014b).

Polar growth in *A. gossypii* requires several microbial structures and their coordination, including polar factors, actin cables, endocytic and secretory vesicles and exocyst factors, which take part in cell wall formation. The actin network is the only cytoskeletal structure involved in *A. gossypii* polar growth, with two to three threads of actin cables present along the hyphal axes, which are the sole responsible for the transport of secretory vesicles (Schmitz and Philippsen 2011).

One particular trait of *A. gosypii* propagation is that septation happens at a different rate of nuclear division, as these phenomena are detached from each other (Kaufmann and Philippsen 2009). This leads to polynucleated hyphal compartments, presenting asynchronous division cycles and variable ploidy (Anderson et al. 2015).

As hyphae and mycelium mature, sporulation starts to occur, affected by the MAP-kinase governed mating pheromone, starvation and filamentous growth responses, and the cAMP/PKA governed morphogenic responses. Hyphal fragmentation occurs at septal sites and sporangia are formed, usually containing eight endospores, which are then released via autolysis for the propagation of the fungus and restarting the cellular cycle (Wendland 2020).

2.2 Carbon and Nitrogen Metabolism

A. gossypii's central nitrogen and carbon metabolism resembles that of most fungi and yeast, with a very close evolutionary and phylogenetic relationship with the yeast *S. cerevisiae*. In fact, over 90% of *A.gossypii* genes present homology and a particular pattern of synteny when compared with the genome of *S. cerevisiae*, with only 5% of the annotated genes in the filamentous fungus having no homologues in the yeast (Dietrich et al. 2004).

As a result, *A gossypii* shows a similar metabolic backbone (Fig. 2.2), including common pathways such as:

- glycolysis and gluconeogenesis,
- fatty acid biosynthesis,
- β-oxidation
- TCA cycle
- Glyoxylate cycle
- Pentose phosphate pathway
- Purine biosynthetic pathway
- Glycine pathway
- Pyrimidine biosynthetic pathway
- Riboflavin synthetic pathway
- Ehrlich pathway
- Mevalonate pathway

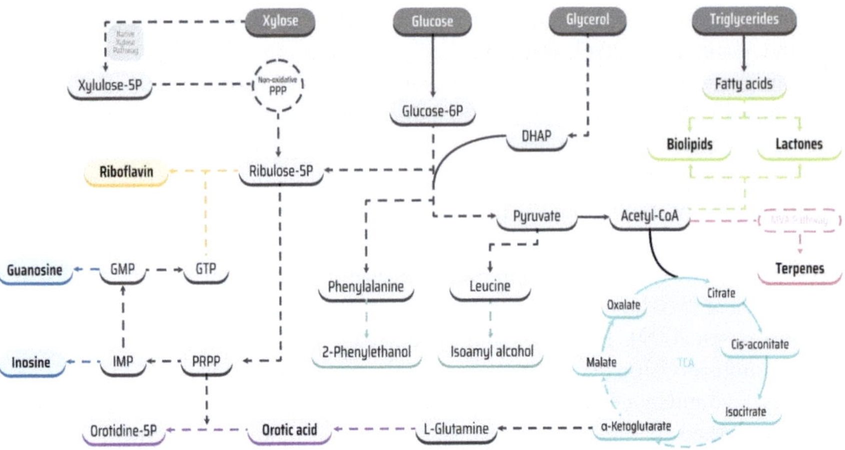

Fig. 2.2 Schematic representation of the main native metabolic pathways in *Ashbya gossypii*. PPP, pentose-phosphate pathway; GMP, guanosine monophosphate; GTP, guanosine triphosphate; IMP, inosine monophosphate; PRPP, phosphoribosyl pyrophosphate; DHAP, dihydroxyacetone 6-phosphate; MVA, mevalonate; TCA, tricarboxylic acid; 5P, 5-phosphate; 6P, 6-phosphate

This fungus is a chemoheterotroph, utilizing mostly organic nutrient sources for growth and energy production. It is also able to use nutrients from inorganic sources, particularly for nitrogen uptake, and minerals as well. Despite its genetic similarity with *S. cerevisiae*, *A. gossypii* still shows several distinctive metabolic traits from this closely related yeast, associated with the presence and absence of specific catalytic functions (Gomes et al. 2014).

A. gossypii can assimilate different and diverse carbon sources. Like most fungi and yeast, it utilizes simple monomeric C6 sugars glucose and fructose through aerobic metabolism and cannot fully utilize C5 sugars like xylose or arabinose for growth (Pridham and Raper 1950). It does possess xylose reductase activity, evidenced by the production of xylitol from xylose (Ribeiro et al. 2012), as well as genes coding for xylose isomerase (*ABR229C*) and xylulose kinase (*AGR324C*), however native expression levels are not sufficient for efficient xylose utilization (Díaz-Fernández et al. 2017). *A. gossypii* is also not able to consume galactose nor lactose, with four enzymes of the Leloir pathway, *GAL1*-coded galactokinase, *GAL7*-coded galactose 1-phosphate uridyltransferase, *GAL10*-coded UDP-glucose epimerase and aldose 1-epimerase, being absent in this organism (Hittinger et al. 2004; Gomes et al. 2014).

A. gossypii is able to consume sugars dimers and more complex polymers. Sucrose is one of these examples, with the fungus owing its ability to hydrolyze sucrose to the secretion of an extracellular invertase encoded by the *AFR529W* (*AgSUC2*) gene (Aguiar et al. 2014a). This fungus has also been reported to oxidize maltose, although at a lower rate than glucose (Mickelson 1950). However, it has been reported that the genes encoding for isomaltase and maltase are absent in this fungus, being the maltose consumption potentially associated with other unknown systems such as 4-alpha-glucanotransferase (Gomes et al. 2014).

Regarding polymeric sugars, *A. gossypii* cannot convert cellobiose nor cellulose (Ribeiro et al. 2012). One aspect to take into consideration is that it doesn't possess β-glucosidase encoding genes, responsible for the conversion of cellobiose to β-D-glucose (Yamada et al. 2013; Gomes et al. 2014). It does have genes coding for a glucan-1,3-β-glucosidase, but this enzyme acts on β-(1->3) and not on the β-(1->4) bonds necessary to consume these substrates (Gomes et al. 2014). On the other hand, *A. gossypii* is able to consume starch, related to having specific enzymatic machinery for this metabolism, namely endopolygalacturonase (which also acts on pectin bonds), α-glucanotransferase and amylo-α-1,6-glucosidase (Ribeiro et al. 2012; Gomes et al. 2014). The fungus can also grow on chitin as sole carbon source, due to having a glucosamine-6-phosphate deaminase that can catalyze its hydrolysis (Gomes et al. 2014).

Apart from sugars, *A. gossypii* can also use other carbon sources, such as glycerol or oils, *e.g.*, corn or soybean, with both substrates being of interest for biotechnological purposes due to their availability and cost (Silva et al. 2022; Martín-González et al. 2025). To use oils as carbon source, *A. gossypii* secretes an extracellular lipase that cleaves triacylglycerides into fatty acids which are then transported to the intracellular space and channeled to a sequential β-oxidation pathway, ultimately resulting in the production of energy and acetyl-CoA (Schwechheimer et al. 2016). Glycerol on its hand is absorbed and metabolized via the glycolysis/gluconeogenesis pathway and converted into acetyl-CoA with pyruvate as intermediary, evidenced by the presence of several syntenic genes to *S. cerevisiae* encoding for glycerol catabolism (Gomes et al. 2014; Aßkamp et al. 2019).

Regarding nitrogen utilization, *A. gossypii* can utilize organic nitrogen in the form of amino acids, as well as inorganic nitrogen. More specifically, in terms of inorganic sources, *A. gossypii* can grow using ammonium as substrate but not from nitrate, with a strong sensitivity to pH that is only seen for inorganic and not in organic nitrogen sources (Ribeiro et al. 2012).

Considering organic nitrogen utilization, some particular aspects have been reported about *A. gossypii* amino acid metabolism. This fungus possesses a primary-amine oxidase which oxidizes a wide range of amines, impacting the metabolism of aminoacids as glycine, serine, threonine, tyrosine and phenylalanine. It also possesses γ-butyrobetaine dioxygenase and trimethyllysine dioxygenase, which impact the hydrolysis of lysine and subsequently the production of carnitine (Gomes et al. 2014). Its ability to metabolize a wide range of nitrogen sources enable *A. gossypii* to grow in a variety of defined and complex substrates such as yeast extract, peptone, casein, animal and corn steep liquor and distillers solubles (Farries and Bell 1930; Wickerham et al. 1946; Tanner et al. 1949; Pfeifer et al. 1950; Smiley et al. 1951). Among these, yeast extract is one of the complex substrates on which there is a deeper understanding of *A. gossypii* metabolism, and potentially one of the most utilized in scientific works. Particularly, yeast extract was shown to be one of the main carbon contributors during growth phase, as shown using ^{13}C labeling, contributing also directly with building blocks for the synthesis of valuable metabolites such as riboflavin (Schwechheimer et al. 2018).

2.3 Metabolites of Biotechnological Interest

2.3.1 *Derivatives of the Purine Pathway: Riboflavin, Nucleosides, and Folates*

Riboflavin, also known as vitamin B2, is probably the main staple in *A. gossypii* biotechnology and was the initial main driver for the biotechnological interest in this fungus. This valuable molecule finds its application in supplementation, impacting energy metabolism and respiration, as well as immune response (Lee et al. 2023). With a growing global market, riboflavin demand was estimated at 9000 t per year in 2015, with 70% being applied as feed and the remaining 30% as food additive and colorant or in pharma (Hoff et al. 2021).

A. *gossypii* stands out as a preferred natural overproducer of this vitamin and is one early example of a biotechnological process overperforming and replacing traditional chemical synthesis. BASF established an industrial process for vitamin riboflavin production based on *A. gossypii* fermentation using vegetable oils as substrate in 1990 and, after 6 years of operation, it enabled the total shutdown of the chemical synthesis production line, outperforming it with a 30% lower emission rate and 43% lower production costs (Wenda et al. 2011; Hoff et al. 2021).

Biosynthesis of riboflavin by *A. gossypii* relies on *de novo* purine metabolism, which provides the guanosine-5-triphosphate (GTP) precursor, and on the pentose-phosphate pathway, that provides the ribulose 5-phosphate (ribulose 5-P) precursor, both channeled into a *de novo* multi-reaction pathway mediated by six RIB genes, with riboflavin as final product (Karos et al. 2004; Gomes et al. 20032014). Upon overproduction, riboflavin accumulates intracellularly, clearly visible by the appearance of crystals inside hyphae or in the fermentation media upon autolysis (Ming et al. ; Schwechheimer et al. 2016).

It is important to note that no enzymes involved in riboflavin biosynthesis were found in *A. gossypii* that could explain its overproducing capacity. This trait is therefore attributed to the regulation of genes within the associated metabolic pathways (Gomes et al. 2014). Riboflavin overproduction is most pronounced at the onset of the stationary phase, when growth declines and sporulation begins (Stahmann et al. 2001). For this reason, riboflavin is considered a pseudo-secondary metabolite (Schlösser et al. 2001) often associated with microbial stress responses such as nutrient depletion or, more importantly, oxidative stress (Schlösser et al. 2007; Kavitha and Chandra 2009; Silva et al. 2019b). In fact, several studies suggest that riboflavin provides protection against oxidative stress, either induced by environmental factors such as UV light, or by reactive oxygen species produced by the plant to fight *A. gossypii*'s parasitic invasion (Walther and Wendland 2012; Silva et al. 2019b).

As a byproduct of the purine metabolism, *A. gossypi* also produces inosine and guanosine, nucleosides that can be applied as sources of umami flavor in the food industry. These are produced from the precursor inosine monophosphate (IMP), an intermediary compound in riboflavin synthesis and the first one in the pathway with a complete flavin ring, generated from PRPP mediated by the *ADE4* gene, which can be converted into other nucleotides precursors like guanosine or adenosine monophosphate and consequently sustaining the production of the corresponding nucleotides and nucleosides (Ledesma-Amaro et al. 2015a).

Folates are other examples of valuable products deriving from the purine biosynthetic pathway. *A. gossypii* is also reported to produce these molecules, with several genes identified as orthologs to *S. cerevisiae* genes involved in folate synthesis, namely *AgFOL1, AgFOL2, AgFOL3,* that code the enzymes responsible for the conversion of GTP into 7,8-dihydrofolate (DHF), which is further converted into tetrahydrofolate (THF) by DHF reductase, encoded by the *AgDFR1* gene (Serrano-Amratian et al. 2016).

2.3.2 *Microbial Lipids*

Fats and oils have a wide range of applications, being found in commodities like food products and fuels as well as in specialty high value products. Due to the growing interest in better performing and more sustainable sources of lipids, several efforts have been made in search of biotechnological alternatives to the traditional sources of lipids like fossil fuels or vegetable oils (Francisco et al. 2023).

Despite not being considered an oleaginous microorganism according to general criteria, *A. gossypii* naturally produces and accumulates lipids (Vorapreeda et al. 2012; Ledesma-Amaro et al. 2014a). It does so using sugars or vegetable oils as substrates, accumulating the produced lipids in intracellular vacuoles (Stahmann et al. 1994) (Fig. 2.3.). Earlier works reporting accumulations between 10 and 20% of dry cell weight (Ledesma-Amaro et al. 2014a) later on increased to 30–40% and even 70% with strain and bioprocess engineering efforts (Ledesma-Amaro et al. 2014a; Díaz-Fernández et al. 2019; Francisco et al. 2023).

Fatty acid and triacylglyceride synthesis rely on central carbon metabolism for the supply of acetyl-CoA which, combined with condensation of malonyl-CoA, sustains lipid production (Ledesma-Amaro et al. 2015b; Lozano-Martínez et al. 2017). A key difference between *A. gossypii* and most conventional oleaginous microorganisms is the absence of genes encoding ATP-citrate lyase, which results in cytosolic acetyl-CoA being supplied exclusively through acetyl-CoA synthase activity (Vorapreeda et al. 2012). On the other hand, *A. gossypii* has important $\Delta 8$ and $\Delta 12$ fatty acid desaturase activities, which are determinant to produce unsaturated fatty acids such as C18:1 and C18:2, differing from its genetically closely related organism, *S. cerevisae,* which relies solely on $\Delta 9$-fatty acid desaturase activity (Ledesma-Amaro et al. 2014b;

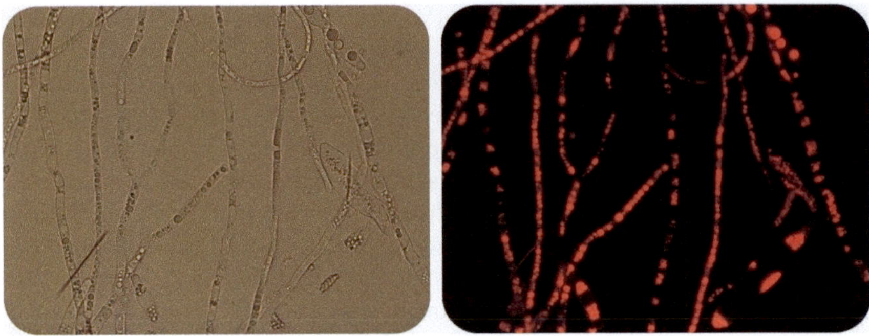

Fig. 2.3 *Ashbya gossypii* hyphae showing intracellular lipid granules visible in bright field microscopy (left panel) and fluorescence microscopy (right panel) using Nile Red for lipid staining

Gomes et al. 2014). Hence, lipids produced by *A. gossypii* have a fatty acid profile dominated by the presence of unsaturated fatty acids, with a major content in oleic acid (C18:1) at around 50% of the total fatty acid content, followed by palmitoleic acid (C16:1) at about 20%, palmitic acid (C16:0) at around 15% and the remaining percentage being composed by residual amounts of other fatty acids, without any of them surpassing 3% of the total amount (Ledesma-Amaro et al. 2014b) (Fig. 2.4).

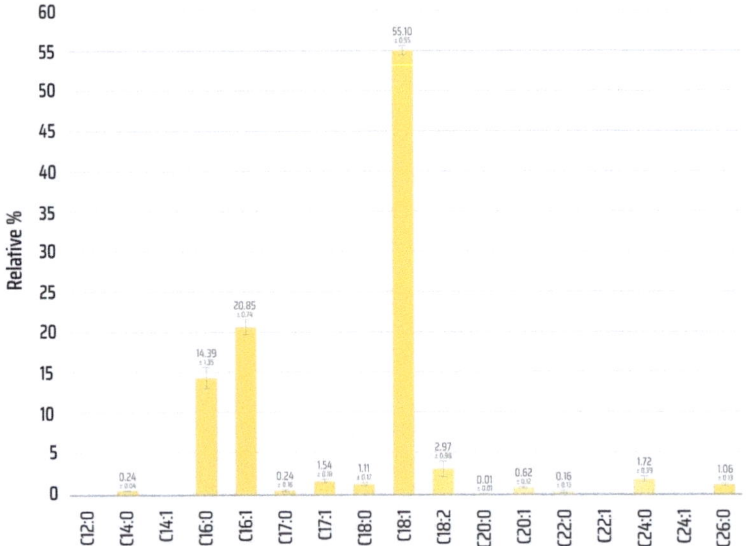

Fig. 2.4 Fatty acid composition of *Ashbya gossypii* microbial lipids. Values extracted from (Ledesma-Amaro et al. 2014b)

2.3.3 *Lactones*

Lactones are another example of metabolites of biotechnological interest synthesized by *A. gossypii*. These molecules have a wide range of applications, often finding their use as fragrances in food and cosmetics but also as drop-in chemicals and building blocks for higher value products (Silva et al. 2021).

A. gossypii stands out for its natural ability to synthesize a wide variety of lactones *de novo* from sugars and, unlike most of the commonly used hosts in bioprocesses, it does not necessarily require the supply of hydroxy fatty acid precursors for lactone production through bioconversion (Silva et al. 2021). Lactone chemical structure is determined by linear chain length and the number of carbons involved in the cycling ring, differentiating γ (5 carbons) and δ (6 carbons) lactones. *A. gossypii* is only known to produce γ-lactones, either saturated (γ-valerolactone, γ-caprolactone, γ-octalactone, γ-nonalactone, γ-undecalactone, γ-dodecalactone, and γ-decalactone), or unsaturated (γ-(Z)-dec-7-enlactone, γ-(E)-dec-5-enlactone and γ-(Z)-dodec-6-enlactone) (Fig. 2.5). Among these, γ-decalactone is one of the main lactones synthesized, being reported to account for approximately 50% of the total lactone content and about 98% of the linear lactones produced (Silva et al. 2019a; Birk et al. 2019).

Ashbya owes this particular trait to the combination of the ability to produce fatty acids with the capacity to perform their *β*-oxidation, therefore resulting in the production of lactones. Fatty acyl-CoA's are produced from acetyl and malonyl-CoA through central carbon metabolism, which are channeled to the *β*-oxidation through the activity of a specific acyl-CoA oxidase system (Silva et al. 2019a). This

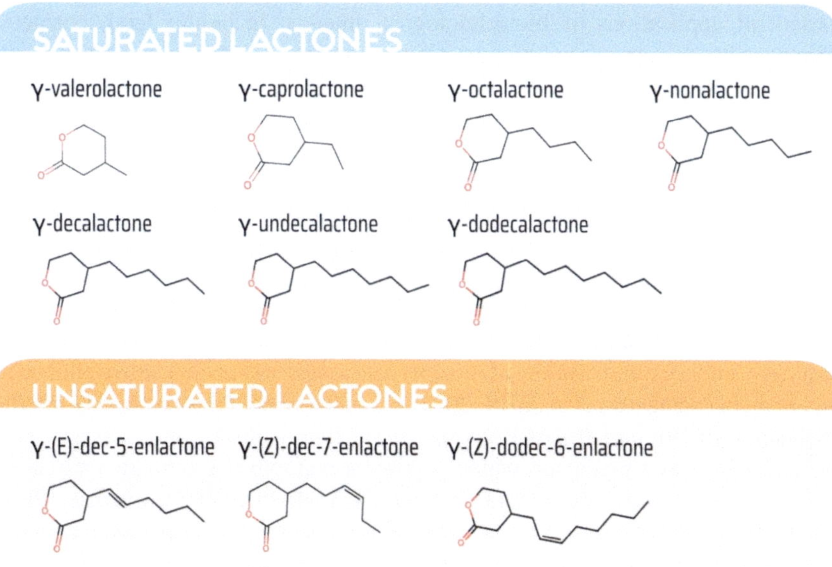

Fig. 2.5 Chemical structures of the saturated and unsaturated lactones produced by *Ashbya gossypii*

system is associated with the *AgPOX1* gene, which is also present in *S. cerevisiae.* It differs from other systems because instead of relying on mitochondrial electron transport chain into receptor-oxygen, the *POX1* associated system transfers electrons directly into oxygen at the peroxisome level (Gomes et al. 2014). However, the *A. gossypii POX1* system is not considered the main distinguishing factor responsible for its unique *de novo* lactone production capacity. In fact, when replacing *A. gossypii POX1* by the *POX2* system from *Yarrowia lipolytica,* lactone production titers remain similar, whereas modifications upstream in the fatty acid production pathway show greater impact on overall lactone production (Silva et al. 2019a). Therefore, current observations indicate that *A. gossypii* lactone biosynthesis is more determined by the production of fatty acyl-CoA precursors, with enzymatic functions linked with desaturation and elongation shown to be determinant for *de novo* production (Silva et al. 2019a). Among this, the desaturase AgDES589p has been proposed as a key contributing enzyme. There are no orthologs of this enzyme found in *S. cerevisiae* nor in other post whole-genome duplication event species of Saccharomycetaceae family, other than *A. gossypii*, which further reinforces the association of this enzymatic function to the observation of the distinctive lactone production capacity of *Ashbya* (Silva et al. 2021).

2.3.4 Other Volatile Organic Compounds: Terpenes and Higher Alcohols

Apart from lactones, terpenes are other aromatic compounds that possess a wide variety of applications of biotechnological interest, including food, cosmetics, pharmaceuticals, agricultural crop protection and fuels (Hoff et al. 2021).

These compounds are a result of the conjugation of isoprene (C5) subunits, with the number of units determining their chemical structure and function, e.g., hemiterpenes (C5), monoterpenes (C10), sesquiterpenes (C15), diterpenes (C20), among others. The precursors for monoterpenes synthesis are isopentenyl diphosphate (IPP) and dimethylallyl diphosphate (DMAPP), which can be obtained via the methylerythritol-4-phosphate (MEP) pathway, most occurring in bacteria and chloroplasts, or via the mevalonate (MVA) pathway, mostly occurring in archaea and eukaryotes (Muñoz-Fernández et al. 2022).

The MVA pathway can also be found in *A. gossypii*, resulting in the ability to naturally produce a wide variety of monoterpenes, such as geraniol, citronellol, nerol, and linalool (Semenova et al. 2019, 2022). As other metabolites previously referred, production of IPP and DMAPP via the mevalonate pathway in *A. gossypii* relies on acetyl-CoA as a precursor, which is transformed into IPP through a cascade of enzymatic controlled reactions and isomerized to DMAPP by IPP isomerase. IPP and DMAPP are then condensed by different prenyl transferases into various prenyl diphosphate precursors which sustain terpene production by terpene synthases (Muñoz-Fernández et al. 2022, 2024).

Along with terpenes and lactones and similarly to other fungi, *A. gossypii* also produces other volatile compounds of aromatic interest, such as alcohols, esters, aldehydes, ketones, volatile acids, pyrazines and others (Ravasio et al. 2014). Among these, a particular focus is given to 3-methyl-1-butanol and 2-phenylethanol, the latter being strongly associated with rose aroma (Ravasio et al. 2014; Birk et al. 2019) at such levels that *A. gossypii* essential oil has been proposed as a viable alternative to rose oil (Semenova et al. 2019). Volatile alcohols are common subproducts of amino acid metabolism via the Ehrlich pathway. Amino acids are converted by a transaminase to α-keto acids, then converted by a decarboxylase into aldehydes and converted to alcohols by a dehydrogenase, with an additional final conversion step into acetate by an acetyltransferase. The amino acid precursor determines the resulting alcohol, e.g., phenylalanine metabolism leads to 2-phenylethanol production whereas leucine originates 3-methyl-1-butanol (Ravasio et al. 2014).

References

Aguiar TQ, Maaheimo H, Heiskanen A, Wiebe MG, Penttilä M, Domingues L (2013) Characterization of the *Ashbya gossypii* secreted *N*-glycome and genomic insights into its *N*-glycosylation pathway. Carbohydr Res 381:19–27. https://doi.org/10.1016/J.CARRES.2013.08.015

Aguiar TQ, Dinis C, Magalhães F, Oliveira C, Wiebe MG, Penttilä M, Domingues L (2014a) Molecular and functional characterization of an invertase secreted by *Ashbya gossypii*. Mol Biotechnol 56:524–534. https://doi.org/10.1007/S12033-013-9726-9

Aguiar TQ, Ribeiro O, Arvas M, Wiebe MG, Penttilä M, Domingues L (2014b) Investigation of protein secretion and secretion stress in *Ashbya gossypii*. BMC Genomics 15:1–18. https://doi.org/10.1186/1471-2164-15-1137

Anderson CA, Roberts S, Zhang H, Kelly CM, Kendall A, Lee C, Gerstenberger J, Koenig AB, Kabeche R, Gladfelter AS (2015) Ploidy variation in multinucleate cells changes under stress. Mol Biol Cell 26:1129–1140. https://doi.org/10.1091/MBC.E14-09-1375

Aßkamp MR, Klein M, Nevoigt E (2019) *Saccharomyces cerevisiae* exhibiting a modified route for uptake and catabolism of glycerol forms significant amounts of ethanol from this carbon source considered as "non-fermentable." Biotechnol Biofuels 12:1–11. https://doi.org/10.1186/S13068-019-1597-2

Ayad-Durieux Y, Knechtle P, Goff S, Dietrich F, Philippsen P (2000) A PAK-like protein kinase is required for maturation of young hyphae and septation in the filamentous ascomycete *Ashbya gossypii*. J Cell Sci 113:4563–4575. https://doi.org/10.1242/JCS.113.24.4563

Birk F, Fraatz MA, Esch P, Heiles S, Pelzer R, Zorn H (2019) Industrial riboflavin fermentation broths represent a diverse source of natural saturated and unsaturated lactones. J Agric Food Chem 67:13460–13469. https://doi.org/10.1021/acs.jafc.9b01154

Díaz-Fernández D, Lozano-Martínez P, Buey RM, Revuelta JL, Jiménez A (2017) Utilization of xylose by engineered strains of *Ashbya gossypii* for the production of microbial oils. Biotechnol Biofuels 10:1–12. https://doi.org/10.1186/S13068-016-0685-9

Díaz-Fernández D, Aguiar TQ, Martín VI, Romaní A, Silva R, Domingues L, Revuelta JL (2019) Microbial lipids from industrial wastes using xylose-utilizing *Ashbya gossypii* strains. Bioresour Technol 293:122054. https://doi.org/10.1016/J.BIORTECH.2019.122054

Dietrich FS, Voegeli S, Brachat S, Lerch A, Gates K, Steiner S, Mohr C, Pöhlmann R, Luedi P, Choi S, Wing RA, Flavier A, Gaffney TD, Philippsen P (2004) The *Ashbya gossypii* genome as a tool for mapping the ancient *Saccharomyces cerevisiae* genome. Science 304:304–307. https://doi.org/10.1126/SCIENCE.1095781

Farries EHM, Bell F (1930) On the metabolism of *Nematospora gossypii* and related fungi, with special reference to the source of nitrogen. Ann Bot 44:423–455. https://doi.org/10.1093/OXF ORDJOURNALS.AOB.A090228

Francisco M, Aguiar TQ, Abreu G, Marques S, Gírio F, Domingues L (2023) Single-cell oil production by engineered *Ashbya gossypii* from non-detoxified lignocellulosic biomass hydrolysate. Fermentation 9:791. https://doi.org/10.3390/FERMENTATION9090791

Gomes D, Aguiar TQ, Dias O, Ferreira EC, Domingues L, Rocha I (2014) Genome-wide metabolic re-annotation of *Ashbya gossypii*: New insights into its metabolism through a comparative analysis with *Saccharomyces cerevisiae* and *Kluyveromyces lactis*. BMC Genomics 15:810. https://doi.org/10.1186/1471-2164-15-810

Hittinger CT, Rokas A, Carroll SB (2004) Parallel inactivation of multiple GAL pathway genes and ecological diversification in yeasts. Proc Natl Acad Sci USA 101:14144–14149. https://doi.org/ 10.1073/PNAS.0404319101

Hoff B, Plassmeier J, Blankschien M, Letzel AC, Kourtz L, Schröder H, Koch W, Zelder O (2021) Unlocking nature's biosynthetic power—metabolic engineering for the fermentative production of chemicals. Angew Chem Int Ed Engl 60:2258–2278. https://doi.org/10.1002/ANIE.202 004248

Karos M, Vilariño C, Bollschweiler C, Revuelta JL (2004) A genome-wide transcription analysis of a fungal riboflavin overproducer. J Biotechnol 113:69–76. https://doi.org/10.1016/J.JBIOTEC. 2004.03.025

Kaufmann A, Philippsen P (2009) Of bars and rings: Hof1-dependent cytokinesis in multiseptated hyphae of *Ashbya gossypii*. Mol Cell Biol 29:771–783. https://doi.org/10.1128/MCB.01150-08

Kavitha S, Chandra TS (2009) Effect of vitamin E and menadione supplementation on riboflavin production and stress parameters in *Ashbya gossypii*. Process Biochem 44:934–938. https://doi. org/10.1016/J.PROCBIO.2009.04.001

Knechtle P, Dietrich F, Philippsen P (2003) Maximal polar growth potential depends on the polarisome component AgSpa2 in the filamentous fungus *Ashbya gossypii*. Mol Biol Cell 14:4140–4154. https://doi.org/10.1091/MBC.E03-03-0167

Ledesma-Amaro R, Santos MA, Jiménez A, Revuelta JL (2014a) Strain design of *Ashbya gossypii* for single-cell oil production. Appl Environ Microbiol 80:1237–1244. https://doi.org/10.1128/ AEM.03560-13

Ledesma-Amaro R, Santos MA, Jiménez A, Revuelta JL (2014b) Tuning single-cell oil production in *Ashbya gossypii* by engineering the elongation and desaturation systems. Biotechnol Bioeng 111:1782–1791. https://doi.org/10.1002/BIT.25245

Ledesma-Amaro R, Buey RM, Revuelta JL (2015a) Increased production of inosine and guanosine by means of metabolic engineering of the purine pathway in *Ashbya gossypii*. Microb Cell Fact 14:1–8. https://doi.org/10.1186/S12934-015-0234-4

Ledesma-Amaro R, Lozano-Martínez P, Jiménez A, Revuelta JL (2015b) Engineering *Ashbya gossypii* for efficient biolipid production. Bioengineered 6:119. https://doi.org/10.1080/216 55979.2015.1011525

Lee TY, Farah N, Chin VK, Lim CW, Chong PP, Basir R, Lim WF, Loo YS (2023) Medicinal benefits, biological, and nanoencapsulation functions of riboflavin with its toxicity profile: a narrative review. Nutr Res 119:1–20. https://doi.org/10.1016/J.NUTRES.2023.08.010

Lozano-Martínez P, Buey RM, Ledesma-Amaro R, Jiménez A, Revuelta JL (2017) Engineering *Ashbya gossypii* strains for *de novo* lipid production using industrial by-products. Microb Biotechnol 10:425–433. https://doi.org/10.1111/1751-7915.12487

Magalhães F, Aguiar TQ, Oliveira C, Domingues L (2014) High-level expression of *Aspergillus niger* β-galactosidase in *Ashbya gossypii*. Biotechnol Prog 30:261–268. https://doi.org/10.1002/ BTPR.1844

Martín-González J, Montero-Bullón JF, Muñoz-Fernández G, Buey RM, Jiménez A (2025) Valorization of waste cooking oil for bioproduction of industrially-relevant metabolites in *Ashbya gossypii*. N Biotechnol 88:32–38. https://doi.org/10.1016/J.NBT.2025.04.002

Mickelson MN (1950) The metabolism of glucose by *Ashbya gossypii*. J Bacteriol 59:659–666. https://doi.org/10.1128/JB.59.5.659-666.1950

Ming H, Lara Pizarro AV, Park EY (2003) Application of waste activated bleaching earth containing rapeseed oil on riboflavin production in the culture of *Ashbya gossypii*. Biotechnol Prog 19:410–417. https://doi.org/10.1021/BP0257325

Muñoz-Fernández G, Martínez-Buey R, Revuelta JL, Jiménez A (2022) Metabolic engineering of *Ashbya gossypii* for limonene production from xylose. Biotechnol Biofuels Bioprod 15:1–13. https://doi.org/10.1186/S13068-022-02176-0

Muñoz-Fernández G, Montero-Bullón JF, Martínez JL, Buey RM, Jiménez A (2024) *Ashbya gossypii* as a versatile platform to produce sabinene from agro-industrial wastes. Fungal Biol Biotechnol 11:1–10. https://doi.org/10.1186/S40694-024-00186-1

Pfeifer VF, Tanner FW, Vojnovich C, Traufler DH (1950) Riboflavin by fermentation with *Ashbya gossypii*. Ind Eng Chem 42:1776–1781. https://doi.org/10.1021/IE50489A027

Pridham TG, Raper KB (1950) *Ashbya gossypii* – Its significance in nature and in the laboratory. Mycologia 42:603–623. https://doi.org/10.1080/00275514.1950.12017863

Ravasio D, Wendland J, Walther A (2014) Major contribution of the Ehrlich pathway for 2-phenylethanol/rose flavor production in *Ashbya gossypii*. FEMS Yeast Res 14:833–844. https://doi.org/10.1111/1567-1364.12172

Ribeiro O, Wiebe M, Ilmén M et al (2010) Expression of *Trichoderma reesei* cellulases CBHI and EGI in *Ashbya gossypii*. Appl Microbiol Biotechnol 87:1437–1446. https://doi.org/10.1007/S00253-010-2610-7

Ribeiro O, Domingues L, Penttilä M, Wiebe MG (2012) Nutritional requirements and strain heterogeneity in *Ashbya gossypii*. J Basic Microbiol 52:582–589. https://doi.org/10.1002/JOBM.201100383

Ribeiro O, Magalhães F, Aguiar TQ, Wiebe MG, Penttilä M, Domingues L (2013) Random and direct mutagenesis to enhance protein secretion in *Ashbya gossypii*. Bioengineered 4:322–331. https://doi.org/10.4161/BIOE.24653

Schlösser T, Schmidt G, Stahmann KP (2001) Transcriptional regulation of 3,4-dihydroxy-2-butanone 4-phosphate synthase. Microbiology 147:3377–3386. https://doi.org/10.1099/00221287-147-12-3377

Schlösser T, Wiesenburg A, Gätgens C et al (2007) Growth stress triggers riboflavin overproduction in *Ashbya gossypii*. Appl Microbiol Biotechnol 76:569–578. https://doi.org/10.1007/S00253-007-1075-9

Schmitz HP, Philippsen P (2011) Evolution of multinucleated *Ashbya gossypii* hyphae from a budding yeast-like ancestor. Fungal Biol 115:557–568. https://doi.org/10.1016/J.FUNBIO.2011.02.015

Schwechheimer SK, Park EY, Revuelta JL, Becker J, Wittmann C (2016) Biotechnology of riboflavin. Appl Microbiol Biotechnol 100:2107–2119. https://doi.org/10.1007/S00253-015-7256-Z

Schwechheimer SK, Becker J, Peyriga L, Portais JC, Wittmann C (2018) Metabolic flux analysis in *Ashbya gossypii* using ^{13}C-labeled yeast extract: Industrial riboflavin production under complex nutrient conditions. Microb Cell Fact 17:162. https://doi.org/10.1186/s12934-018-1003-y

Semenova E, Presniakova V, Kozlovskaya V, Markelova N, Gusev A, Linert W, Kurakov A, Shpichka A (2022) The *in vitro* cytotoxicity of *Eremothecium* oil and its components—aromatic and acyclic monoterpene alcohols. Int J Mol Sci 23:3364. https://doi.org/10.3390/ijms23063364

Semenova EF, Presnyakova EV, Shpichka AI, Presnyakova VS (2019) *Eremothecium* oil biotechnology as a novel technology for the modern essential oil production. In: Malik S (ed) Essential oil research: trends in biosynthesis, analytics, industrial applications and biotechnological production. Springer, Cham, pp 401–435. https://doi.org/10.1007/978-3-030-16546-8_15

Serrano-Amatriain C, Ledesma-Amaro R, López-Nicolás R, Ros G, Jiménez A, Revuelta JL (2016) Folic acid production by engineered *Ashbya gossypii*. Metab Eng 38:473–482. https://doi.org/10.1016/J.YMBEN.2016.10.011

Silva R, Aguiar TQ, Coelho E et al (2019a) Metabolic engineering of *Ashbya gossypii* for deciphering the *de novo* biosynthesis of γ-lactones. Microb Cell Fact 18:62. https://doi.org/10.1186/s12934-019-1113-1

Silva R, Aguiar TQ, Oliveira R, Domingues L (2019b) Light exposure during growth increases riboflavin production, reactive oxygen species accumulation and DNA damage in *Ashbya gossypii* riboflavin-overproducing strains. FEMS Yeast Res 19:1. https://doi.org/10.1093/femsyr/foy114

Silva R, Coelho E, Aguiar TQ (2021) Domingues L (2021) Microbial biosynthesis of lactones: gaps and opportunities towards sustainable production. Appl Sci 11:8500. https://doi.org/10.3390/APP11188500

Silva R, Aguiar TQ, Domingues L (2022) Orotic acid production from crude glycerol by engineered *Ashbya gossypii*. Bioresour Technol Rep 17:100992. https://doi.org/10.1016/j.biteb.2022.100992

Smiley KL, Sobolov M, Austin FL, Rasmussen RA, Smith MB, van Lanen JM, Stone L, Boruff CS (1951) Biosynthesis of riboflavin, *L. bulgaricus* factor, and other growth factors. Laboratory and pilot plant studies of biosynthesis by *A. gossypii* cultivated on grain stillage media. Ind Eng Chem 43:1380–1384. https://doi.org/10.1021/IE50498A033

Stahmann KP, Kupp C, Feldmann SD, Sahm H (1994) Formation and degradation of lipid bodies found in the riboflavin-producing fungus *Ashbya gossypii*. Appl Microbiol Biotechnol 42:121–127. https://doi.org/10.1007/BF00170234/METRICS

Stahmann KP, Arst HN, Althöfer H, Revuelta JL, Monschau N, Schlüpen C, Gätgens C, Wiesenburg A, Schlösser T (2001) Riboflavin, overproduced during sporulation of *Ashbya gossypii*, protects its hyaline spores against ultraviolet light. Environ Microbiol 3:545–550. https://doi.org/10.1046/J.1462-2920.2001.00225.X

Tanner FW, Vojnovich C, Van Lanen JM (1949) Factors affecting riboflavin production by *Ashbya gossypii*. J Bacteriol 58:737–745. https://doi.org/10.1128/JB.58.6.737-745.1949

Vorapreeda T, Thammarongtham C, Cheevadhanarak S, Laoteng K (2012) Alternative routes of acetyl-CoA synthesis identified by comparative genomic analysis: involvement in the lipid production of oleaginous yeast and fungi. Microbiology 158:217–228. https://doi.org/10.1099/MIC.0.051946-0

Walther A, Wendland J (2012) Yap1-dependent oxidative stress response provides a link to riboflavin production in *Ashbya gossypii*. Fungal Genet Biol 49:697–707. https://doi.org/10.1016/J.FGB.2012.06.006

Wenda S, Illner S, Mell A, Kragl U (2011) Industrial biotechnology—the future of green chemistry? Green Chem 13:3007–3047. https://doi.org/10.1039/C1GC15579B

Wendland J (2020) Sporulation in *Ashbya gossypii*. J Fungi (Basel) 6:157. https://doi.org/10.3390/JOF6030157

Wickerham LJ, Flickinger MH, Johnston RM (1946) The production of riboflavin by *Ashbya gossypii*. Arch Biochem 9:95–98

Yamada R, Nakatani Y, Ogino C, Kondo A (2013) Efficient direct ethanol production from cellulose by cellulase- and cellodextrin transporter-co-expressing *Saccharomyces cerevisiae*. AMB Express 3:1–7. https://doi.org/10.1186/2191-0855-3-34

Chapter 3
Optimization of the *Ashbya gossypii* Natural Metabolism for Biotechnological Applications

Abstract *Ashbya gossypii* has emerged as a promising microbial platform for the sustainable production of high-value biochemicals. This chapter explores the optimization of its natural metabolism through the development and application of an increasingly sophisticated molecular toolbox. Core components of this toolbox include efficient transformation protocols, integrative and plasmid-based expression systems, a library of well-characterized promoters and terminators, scarless genome editing methods, and modular platforms for assembling multigene expression cassettes. Combined with the *A. gossypii*'s inherent metabolic potential, compact and well-annotated genome, efficient homologous recombination system, and close genetic relationship to *Saccharomyces cerevisiae*, these tools have enabled the implementation of advanced metabolic engineering and synthetic biology strategies. Such strategies have been employed to rewire endogenous pathways, enhance precursor supply, expand substrate utilization, and boost overall biosynthetic capacity. As a result, *A. gossypii* has been successfully engineered for the production of a wide array of valuable compounds, including riboflavin, nucleosides, folates, orotic acid, lipids, lactones, monoterpenes, gangliosides, and recombinant proteins. This chapter provides a comprehensive overview of the molecular and metabolic engineering approaches that have transformed *A. gossypii* into a versatile and industrially relevant chassis for next-generation microbial manufacturing.

Keywords *Ashbya gossypii* · Molecular toolbox · Genome editing · Metabolic engineering · Synthetic biology · Microbial chassis · Biotechnological production

3.1 Molecular Toolbox

The transformation of *A. gossypii* into a robust microbial platform has been underpinned by the development of a diverse and evolving molecular toolbox, which has enabled the genetic optimization of its natural metabolism toward improved production of biotechnologically relevant compounds. This toolbox includes classical mutagenesis techniques and rational genetic engineering tools, namely efficient

© The Author(s), under exclusive license to Springer Nature Switzerland AG 2026
T. Q. Aguiar et al., *Ashbya gossypii in Biotechnology*,
SpringerBriefs in Molecular Science, https://doi.org/10.1007/978-3-032-12435-7_3

Fig. 3.1 Overview of the *Ashbya gossypii* molecular toolbox

transformation protocols, plasmid-based and integrational expression systems, well-characterized promoter and terminator sequences, targeted gene deletion/insertion tools, scarless genome editing methods, and versatile tools for rapid modular assembly of multigene integrative cassettes from available promoter, terminator, and selectable marker bio-bricks (Fig. 3.1).

3.1.1 Random Mutagenesis

Historically, strain improvement in *A. gossypii* began with classical random mutagenesis. Physical mutagens like UV radiation (Pridham and Raper 1952; Perlman 1979; Schmidt et al. 1996; Park et al. 2007; Wei et al. 2012; Dessipri et al. 2022) and chemical agents such as nitrogen mustard (Pridham and Raper 1952), acridine half-mustard ICR 191 (Lago and Kaplan 1981), N-Nitroso-N-methylurea (NMU) (Lago and Kaplan 1981), ethyl methanesulfonate (EMS) (Lago and Kaplan 1981; Ribeiro et al. 2013), and N-methyl-N'-nitro-N-nitrosoguanidine (MNNG) (Lizama et al. 2007; Tajima et al. 2009) were commonly used to induce genetic diversity, followed by phenotypic screening for desirable traits. Adaptive evolution under selective pressure has also been effective in enriching for spontaneous mutants with improved phenotypes (Pfeifer et al. 1950; Lago and Kaplan 1981; Sugimoto et al. 2010).

Beyond these traditional approaches, alternative random mutagenesis techniques have been explored, including insertional mutagenesis via *in vitro* transposition (Santos et al. 2005) and disparity mutagenesis using error-prone DNA polymerases (Park et al. 2011). However, classical genetics in *A. gossypii* faces significant limitations, most notably the absence of a known sexual cycle, which precludes mating-based genetic analyses (Wendland et al. 2011; Dietrich et al. 2013). Additionally, spore clustering via their terminal filaments complicates the isolation of single spores for screening (Wendland et al. 2011).

3.1.2 Transformation Protocols, Replicative Plasmids, and Gene Targeting

The advent of recombinant DNA techniques significantly broadened the genetic engineering possibilities in *A. gossypii*. In 1991, using a PEG-mediated protoplast transformation method, Wright and Philippsen (1991) first transformed *A. gossypii* and demonstrated that plasmids carrying *S. cerevisiae* autonomously replicating sequences (*ARS1* and *2μ*) replicate in *A. gossypii*, enabling episomal gene expression. In 1994, Steiner and Philippsen (1994) further improved the efficiency of this transformation method by expressing the previously used kanamycin resistance gene (*kan*) of the *Escherichia coli* transposon Tn*903* under the control of the newly characterized *A. gossypii TEF* promoter (P_{AgTEF}), which enabled improved selection of G418/geneticin-resistant transformants. In 1996, building on the recently discovered homologous recombination efficiency of *A. gossypii* (Steiner et al. 1995), Altmann-Jöhl and Philippsen (1996) successfully generated threonine auxotrophic strains by targeted gene disruption. Disruption and deletion cassettes for the *AgTHR4* gene were constructed using the *kanMX* selectable marker, comprising the *kan* gene under the control of the *A. gossypii TEF* promoter and terminator (Wach et al. 1994), or modified versions thereof, flanked by upstream and downstream sequences with > 900 bp homology to the *AgTHR4* locus (Altmann-Jöhl and Philippsen 1996). Using a similar strategy, an auxotrophic strain deleted for both the *AgLEU2* and *AgTHR4* genes was also generated, whose auxotrophies could be complemented by selection markers comprising either the native or the corresponding *S. cerevisiae* homologs (Altmann-Jöhl and Philippsen 1996; Ayad-Durieux et al. 2000). Around 1998, Revuelta established an electroporation transformation protocol for gene integration into *A. gossypii* (Althöfer et al. 1999; Monschau et al. 1998), which was subsequently optimized by Wendland et al. (2000) to enable PCR-based gene targeting using homology arms as short as 40 base pairs.

These integrative transformation systems generate heterokaryotic transformants that carry both wild and mutant nuclei in one mycelium, which upon sporulation under selective conditions produce uninucleate and haploid spores from which homokaryotic mycelia solely carrying mutant nuclei can be readily isolated (Steiner et al. 1995; Wendland et al. 2000). Following clonal selection, genomic integration yields homokaryotic strains that remain stable under non-selective conditions, in contrast to plasmid-based systems, which require continuous selective pressure, as plasmids can otherwise be lost within a single spore germination cycle (Aguiar et al. 2014a; Muñoz-Fernández et al. 2020). Taken together, these works laid the foundation for targeted metabolic engineering, by enabling precise gene disruption and stable overexpression in *A. gossypii* (Aguiar et al. 2015).

3.1.3 Selectable Markers and Marker Recycling

Selectable markers enable the distinction between transformed and untransformed mycelia, playing a crucial role in plasmid maintenance and in the clonal selection of homokaryotic mutants. They generally are classified into dominant drug-resistance markers, auxotrophic markers and bidirectional/counterselectable markers (Oliveira et al. 2017). Auxotrophic and bidirectional/counterselectable markers can only be used in mutant strains with adequate genetic backgrounds. In contrast, dominant drug-resistance markers can be used in any strain that is sensitive to the selective agent, making them the most commonly used.

As previously mentioned, auxotrophic markers comprising the *AgTHR4*, *AgLEU2* or *ScLEU2* genes with their own promoters and terminators have been successfully used for selection of *A. gossypii* auxotrophic mutants (Altmann-Jöhl and Philippsen 1996; Ayad-Durieux et al. 2000; Alberti-Segui et al. 2001; Kaufmann 2009). *A. gossypii* strains auxotrophic to uridine/uracil (*Agura3*) (Aguiar et al. 2014a; Jiménez et al. 2020), adenine (*Agade1*, *Agade2* and *Agade4*) (Aguiar et al. 2014a; Jiménez et al. 2005, 2019, 2020), histidine (*Aghis3*) (Jiménez et al. 2020) and tryptophan (*Agtrp1*) (Jiménez et al. 2020) have also been generated, which are suitable for future complementation-based selection studies, namely using bidirectional/counterselectable markers such as the *AgURA3* or *ScURA3* genes conferring uridine/uracil prototrophy and 5-FOA sensitivity (Aguiar et al. 2014a).

While developing PCR-based gene targeting, Wendland et al. (2000) constructed a modified version of the *kanMX* drug-resistance marker comprising the promoter and terminator regions of the *S. cerevisiae TEF2* in substitution of those of the *A. gossypii TEF* (*GEN3* marker) to avoid undesired integration into the *A. gossypii TEF* locus. Other fully heterologous drug-resistance markers have been subsequently developed to support genetic manipulation in *A. gossypii*, including the *NAT5* or *NATPS* (Dünkler and Wendland 2007; Kaufmann 2009) and the *BLE3* (Ribeiro et al. 2013; Aguiar et al. 2014a) markers, conferring resistance to clonNAT/nourseothricin and phleomycin, respectively. Still, hygromycin B (Schlüpen et al. 2003; Mateos et al. 2006; Jiménez et al. 2008; Serrano-Amatriain et al. 2016), G418/geneticin (Jiménez et al. 2005; Ledesma-Amaro et al. 2018; Muñoz-Fernández et al. 2024), and clonNAT/nourseothricin (Ledesma-Amaro et al. 2014a; 2018) resistance genes flanked by homologous *A. gossypii* promoter and terminator sequences have continued to be widely used for dominant selection of *A. gossypii* transformants.

Selectable markers are not abundant in variety and are undesirable for industrial purposes (Oliveira et al. 2017). With the adaptation of the Cre-*lox*P recombination system to *A. gossypii* (Aguiar et al. 2014a), marker recycling and multi-step genome engineering without accumulation of resistance genes became possible from 2014 on. Since then, the above markers flanked by *lox*P inverted repeat sequences and plasmids with different selectable markers for Cre recombinase expression have been made available and used for the edition of the *A. gossypii* genome (Aguiar et al. 2014a; Ledesma-Amaro et al. 2014a; 2018; Serrano-Amatriain et al. 2016). These markers and plasmids significantly expanded the *A. gossypii* genetic toolkit, enabling more flexible and efficient engineering strategies.

3.1.4 Promoters, Terminators, and Reporter Proteins/Genes

Promoters driving high gene expression in *A. gossypii* were early discovered during the 1990s. These include the strong constitutive promoters from the *A. gossypii* translation elongation factor 1α (P_{AgTEF}) and glyceraldehyde-3-phosphate dehydrogenase (P_{AgGPD}) genes (Steiner and Philippsen 1994; Revuelta et al. 1999). These promoters support very high expression of both homologous and heterologous genes in *A. gossypii*, and have been of major importance for the genetic and metabolic engineering of this fungus (Monschau et al. 1998; Maeting et al. 1999; Wendland 2003; Dünkler and Wendland 2007; Kaufman 2009; Walther and Wendland 2012; Ledesma-Amaro et al. 2014a; 2018; Magalhães et al. 2014; Silva et al. 2019a; Muñoz-Fernández et al. 2022).

In addition to these widely used promoters, several endogenous and *S. cerevisiae* constitutive promoters have been validated for use in *A. gossypii* throughout the years, including P_{ScTEF2} (Wendland et al. 2000), P_{ScTEF1} (Wendland et al. 2011; Lengeler et al. 2013), P_{LEU2} from both *A. gossypii* (Schlüpen 2003) and *S. cerevisiae* (Kaufmann 2009), P_{AgRHO1} (Bauer et al. 2004; Nordmann et al. 2014), P_{ScHIS3} (Helfer and Gladfelter 2006; Kaufmann 2009; Gibeaux et al. 2013), P_{ScPDC1} (Kaufmann 2009), P_{AgCTS2} (Dünkler et al. 2008; Grünler et al. 2010), P_{AgFIG2} (Grünler et al. 2010), P_{PGK1} from both *S. cerevisiae* (Ribeiro et al. 2010; Magalhães et al. 2014) and *A. gossypii* (Ledesma-Amaro et al. 2018), P_{ScADH1} (Magalhães et al. 2014), P_{AgRIB1} (Walther and Wendland 2012; Wasserstrom et al. 2015), P_{AgRIB7} (Ledesma-Amaro et al. 2015a; 2015b), P_{AgACT1} (Wasserstrom et al. 2017), and P_{AgFBA1} (Ledesma-Amaro et al. 2018). Of these, P_{AgRIB7} has been used for gene underexpression in *A. gossypii* (Ledesma-Amaro et al. 2015a; 2015b; Serrano-Amatriain et al. 2016), and P_{AgFIG2} and P_{AgCTS2} were demonstrated to be regulated by Tec1 (Grünler et al. 2010).

Conditional metabolic expression can also be obtained using regulatable promoters such as P_{AgRIB3} (Schlösser et al. 2001, 2007), P_{AgRIB4} (Walther and Wendland 2012; Wasserstrom et al. 2017), P_{AgRIB5} (Wasserstrom et al. 2017), P_{AgYAP1} (Walther and Wendland 2012), P_{AgIME1} (Wasserstrom et al. 2013; Wasserstrom et al.

2017), P_{AgICL1} (Maeting et al. 1999; Lozano-Martínez et al. 2016), P_{MET3} from both *S. cerevisiae* and *A. gossypii* (Dünkler and Wendland 2007; Kaufmann 2009), $P_{ScTHI13}$ (Kaufmann 2009), P_{AgSUC2} (Aguiar et al. 2014b), and P_{ScADH2} (Aguiar et al. 2014a). Among these, P_{AgICL1}, P_{MET3} from both *S. cerevisiae* and *A. gossypii*, $P_{ScTHI13}$, P_{AgSUC2}, and P_{ScADH2} are repressible, and the other are auto-inducible. Noteworthy, P_{AgRIB4} drives high basal gene expression levels (Walther and Wendland 2012; Ledesma-Amaro et al. 2015a), which can be further induced by oxidative stress or nutritionally challenging conditions (Walther and Wendland 2012; Wasserstrom et al. 2017).

A recent systematic characterization of several endogenous *A. gossypii* promoters using the *Renilla*-Firefly dual luciferase reporter system further expanded this catalogue, providing additional tools for precise gene expression modulation (Muñoz-Fernández et al. 2021): three new strong promoters for gene overexpression, $P_{AgCCW12}$, P_{AgSED1}, and P_{AgTSA1}, of which P_{AgSED1} drives particularly high gene overexpression in lipid-based culture media. Additionally, seven novel promoters, $P_{AgHSP26}$, $P_{AgAGL366C}$, $P_{AgTMA10}$, P_{AgCWP1}, $P_{AgAFR038W}$, P_{AgPFS1}, and P_{AgCDA2}, also became available for gene underexpression, of which $P_{AgHSP26}$, $P_{AgAGL366C}$, $P_{AgTMA10}$, and P_{AgPFS1} display lipid-dependent regulatory properties (Muñoz-Fernández et al. 2021) (Table 3.1).

Several transcription terminator sequences have been used together with the above-mentioned promoters to enable proper mRNA termination, crucial for the full-length expression of functional proteins. These include T_{AgTEF} (Monschau et al. 1998; Jiménez et al. 2005), T_{ScTEF2} (Wendland et al. 2000), T_{ScPDC1} (Kaufmann 2009), T_{ScLYS2} (Kaufmann 2009), T_{ScURA3} (Knechtle et al. 2003; Hungerbuehler et al. 2007; Kaufmann 2009; Lee et al. 2015), T_{LEU2} from both *S. cerevisiae* (Kaufmann 2009) and *A. gossypii* (Schlüpen 2003), T_{PGK1} from both *S. cerevisiae* (Ribeiro et al. 2010; Magalhães et al. 2014; Aguiar et al. 2014b) and *A. gossypii* (Ledesma-Amaro et al. 2018; Muñoz-Fernández et al. 2021), T_{CYC1} from both *S. cerevisiae* (Aguiar et al. 2014a; Jiménez et al 2019) and *A. gossypii* (Ledesma-Amaro et al. 2018), T_{AgENO1} (Ledesma-Amaro et al. 2018; Jiménez et al 2020; Muñoz-Fernández et al. 2022), T_{AgICL1} (Ledesma-Amaro et al. 2018) and T_{AgRIP1} (Ledesma-Amaro et al. 2018). Beyond all of the above RNA polymerase II promoters and terminators, the RNA polymerase III promoter and terminator sequences from the *AgSNR52* were demonstrated to drive the expression of sgRNA in *A. gossypii* (Jiménez et al. 2019, 2020).

To study gene expression, other reporter proteins have been employed in addition to the above-mentioned *Renilla*-Firefly dual luciferase reporter system, including β-galactosidase from *E. coli* (Steiner and Philippsen 1994; Schlösser et al. 2001; 2007), *Streptococcus thermophilus* (Dünkler and Wendland 2007; Dünkler et al. 2008; Grünler et al. 2010; Walther and Wendland 2012) and *Aspergillus niger* (Magalhães et al. 2014), as well as green fluorescent protein (GFP) (Dünkler and Wendland 2007; Kaufmann 2009). GFP and other (cyan, green/yellow, yellow, and red) fluorescent proteins have also been widely used in *A. gossypii* as reporters to study protein subcellular localization, together with different fluorescent dyes used to stain organelles and cytoskeletal elements (Ayad-Durieux et al. 2000; Alberti-Segui et al.

Table 3.1 Comparative strength of different promoters in *Ashbya gossypii*, assessed using reporter protein assays under standardized conditions

Promoter name	Promoter length (bp)	Promoter activity	Reference
		Relative to P_{AgGPD}	
$P_{AgCCW12}$	432	4.96	
P_{AgSED1}	711	4.50	
P_{AgTSA1}	255	3.33	
P_{AgGPD}	373	1.00	
$P_{AgHSP26}$	1000	0.24	Muñoz-Fernández et al. (2021)
$P_{AgAGL366C}$	672	0.11	
$P_{AgTMA10}$	258	0.11	
P_{AgCWP1}	950	0.05	
$P_{AgAFR038W}$	202	0.03	
P_{AgPFS1}	176	0.02	
P_{AgCDA2}	325	0.02	
		Relative to P_{AgGPD}	
P_{AgTEF}	388	2.24	
P_{AgGPD}	371	1.00	Magalhães et al. (2014)
P_{ScPGK1}	576	0.29	
P_{ScADH1}	697	0.29	
		Relative to P_{AgTEF}	
P_{AgTEF}	388	1.00	Dünkler and Wendland (2007)
P_{AgMET3}	269	0.71 to 0[a]	
		Relative to P_{AgRIB4}	
P_{AgRIB4}	576	1.00 to 2.00[b]	
P_{AgRIB5}	549	0.24	Walther and Wendland (2012)
P_{AgRIB1}	549	0.13	
P_{AgRIB3}	621	0.10	

Relative promoter activity was determined based on reporter expression levels, enabling direct comparison of transcriptional strength. Color coding from red to green indicates promoter strength, with red representing strong, yellow medium, and green weak activity

[a] After induction with 3 mM H_2O_2

[b] After repression with 2.5 mM Methionine

2001; Wendland 2003; Knechtle et al. 2003; Bauer et al. 2004; Gladfelter et al. 2006; Hungerbuehler et al. 2007; Kaufmann 2009; DeMay et al. 2010; Lang et al. 2010; Grava and Philippsen 2010; Finlayson et al. 2011; Gerstenberger et al. 2012; Lee et al. 2015; Wabner et al. 2019). Moreover, given the characteristic red pigmentation phenotype of *Adade1* and *Agade2* mutants due to the accumulation of intermediates of the purine pathway, both *AgADE1* and *AgADE2* have been used as reporter genes (Aguiar et al. 2014a; Jiménez et al. 2019).

3.1.5 *Modular Genetic Toolkits and CRISPR/Cas Genome Editing*

The development of modular genetic toolkits and marker-less genome editing platforms has significantly expanded the *A. gossypii* molecular toolbox. A notable advancement was the establishment of a Golden Gate-based modular toolkit tailored specifically for *A. gossypii*, which allows rapid one-step assembly of expression and editing cassettes using standardized genetic parts (Ledesma-Amaro et al. 2018; Muñoz-Fernández et al. 2021). This system provides interchangeable modules for promoters, coding sequences, terminators, and selectable markers, enabling combinatorial pathway construction and rapid prototyping. Although originally designed for yeast-like organisms, it has been successfully adapted for *A. gossypii*, streamlining the design-build-test cycle for metabolic engineering (Ledesma-Amaro et al. 2018; Muñoz-Fernández et al. 2022). Its compatibility with existing yeast vectors further enhances its plug-and-play functionality and broadens its application scope.

Complementing this toolkit, clustered regularly interspaced short palindromic repeats (CRISPR)-Cas-mediated marker-less genome editing has emerged as a powerful addition to the *A. gossypii* genetic toolbox. Jiménez et al. (2019) developed a one-vector CRISPR/Cas9 system containing all necessary components for genome editing in *A. gossypii:* Cas9 endonuclease, single guide RNA (sgRNA), and donor DNA (dDNA) repair template. This system leverages the *A. gossypii*'s highly efficient homologous recombination machinery to enable precise, marker-free, and scarless gene deletions, disruptions, and point mutations (Jiménez et al. 2019; Díaz-Fernández et al. 2020; Muñoz-Fernández et al. 2020). By co-delivering Cas9, sgRNA targeting canonical 5′-NGG-3′ protospacer adjacent motif (PAM) sequences, and dDNA within a single plasmid, editing efficiencies ranging 36–80% can be achieved (Jiménez et al. 2019).

Building on this system, a one-vector CRISPR/Cpf1 system was later developed, in which *Streptococcus pyogenes* Cas9 and its associated sgRNA were substituted by the *Lachnospiraceae bacterium* Cpf1 (recently renamed as Cas12a) and corresponding CRISPR RNA (crRNA) targeting 5′-TTTN-3′ PAM sequences. This expanded the range of editable genomic *loci* by overcoming PAM sequence constraints and enabled both single and multiplex genome editing of up to four genes, with reported efficiencies of 19–77% for single edits and 10–30% for multiplex edits (Jiménez et al. 2020).

Together, these modular DNA assembly and CRISPR/Cas-based genome editing toolkits represent a significant advance for *A. gossypii* synthetic biology, facilitating iterative genome modifications and metabolic pathway optimization.

3.2 Metabolic Engineering and Synthetic Biology Strategies

With a native biosynthetic capacity well-suited for the production of various biotechnologically valuable compounds, *A. gossypii* has been increasingly harnessed as a microbial cell factory through targeted metabolic modifications. Its favorable genetic traits, including efficient homologous recombination, genetic tractability and stability, a compact and well-annotated genome, and close genetic similarity to *S. cerevisiae,* combined with the ongoing development of a versatile molecular toolbox, have enabled advanced metabolic engineering and positioned *A. gossypii* as a promising chassis for synthetic biology applications. Accordingly, synthetic and metabolic engineering strategies have been applied to rewire endogenous pathways, enhance precursor supply, eliminate competing or regulatory bottlenecks, and optimize or expand substrate utilization. These efforts have significantly broadened the metabolic potential of *A. gossypii*, establishing it as a robust host for the production of high-value biochemicals under industrially relevant conditions.

3.2.1 *Pentose Phosphate, Purine, Riboflavin and Folate Biosynthesis Pathways*

3.2.1.1 Riboflavin

To enhance riboflavin production in *A. gossypii*, both classical strain improvement and rational metabolic engineering approaches have been successfully employed (Revuelta et al. 2017). The advent of recombinant DNA technology was pivotal in enabling the rational engineering of *A. gossypii*, as it led to the development of a comprehensive molecular toolbox that facilitated precise genetic manipulation and allowed for the detailed molecular characterization of the riboflavin biosynthetic genes, as well as other key genes involved in riboflavin production (Revuelta et al. 2017).

Over the years, rational metabolic engineering strategies targeting the improvement of riboflavin production by *A. gossypii* have focused on increasing precursor availability, optimizing the metabolic flux through the riboflavin biosynthetic pathway, and enhancing riboflavin secretion.

A key target has been the purine biosynthetic pathway, since guanosine triphosphate (GTP) serves as a critical precursor for riboflavin biosynthesis (Fig. 3.2). Overexpression of feedback-insensitive variants of phosphoribosyl pyrophosphate (PRPP) amidotransferase (AgAde4^{V,K,W}) (Jiménez et al. 2005) and PRPP synthetases (AgPrs2,4^{I,Q} and AgPrs3^{I,Q}) (Jiménez et al. 2008) led to increases in riboflavin production of up to tenfold and twofold, respectively. Deletion of *AgURA3*, which blocks the *de novo* pyrimidine biosynthesis, was also shown to redirect the PRPP flux toward purine biosynthesis, thus enhancing precursor supply (Silva et al. 2015, 2019b). Metabolic modeling further guided the deletion/underexpression of *AgADE12* (adenylosuccinate synthase), which diverts the flux of the purine biosynthesis pathway from adenosine monophosphate (AMP) to the guanosine monophosphate (GMP) and ultimately GTP, leading to a 2.5-fold increase in riboflavin production (Ledesma-Amaro et al. 2015a). Additionally, deregulation of AgBas1, a transcription factor controlling the purine biosynthesis pathway, via deletion of its regulatory domain, led to constitutive expression of its target genes and a tenfold increase in riboflavin production (Mateos et al. 2006).

Another key strategy has been increasing intracellular glycine availability, another essential precursor for purine biosynthesis (Fig. 3.2). Three complementary approaches were shown to be effective in increasing glycine pools and thereby enhancing riboflavin production: overexpression of *AgGLY1* (threonine aldolase)

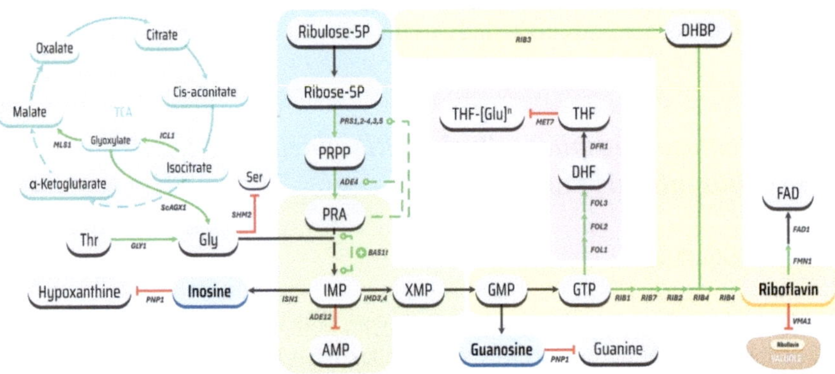

Fig. 3.2 Schematic representation of the interconnected pentose phosphate (blue), purine (green), riboflavin (orange), and folate (pink) biosynthesis pathways in *Ashbya gossypii,* highlighting shared precursors, branch points, and key metabolic engineering targets for enhanced production of riboflavin, flavin adenine dinucleotide (FAD), purine nucleosides, and folates. Green arrows indicate gene overexpression; red lines indicate gene knockout or underexpression; green broken lines indicate abolished end-product inhibition mechanism and transcription activation of the purine biosynthesis genes by the truncated transcription factor AgBas1t. AMP, adenosine monophosphate; DHBP, 3,4-dihydroxy-2-butanone-4-phosphate; DHF, 7,8-dihydrofolate; Gly, glycine; GMP, guanosine monophosphate; GTP, guanosine triphosphate; IMP, inosine monophosphate; PRA, 5-phosphoribosylamine; PRPP, phosphoribosylpyrophosphate; ribose-5P, ribose 5-phosphate; ribulose-5P, ribulose 5-phosphate; Ser, serine; THF, tetrahydrofolate; THF-[Glu]^n, tetrahydrofolyl-polyglutamate; Thr, threonine; XMP, xanthosine monophosphate

(Monschau et al. 1998), deletion of *AgSHM2* (serine hydroxymethyltransferase) (Schlüpen et al. 2003), and heterologous expression of the *S. cerevisiae ScAGX1* (alanine:glyoxylate aminotransferase) (Kato and Park 2006).

Direct engineering of the riboflavin biosynthetic pathway itself has also yielded substantial improvements (Althöefer and Revuelta 2003; Ledesma-Amaro et al. 2015a). Systematic overexpression of *AgRIB* genes revealed that *AgRIB1* and *AgRIB3* are rate-limiting. Co-overexpression of *AgRIB1*, *AgRIB2*, *AgRIB3*, *AgRIB5* and *AgRIB7* led to a threefold increase in riboflavin production, which combined with the underexpression of *AgADE12* resulted in a fivefold increase in riboflavin production, with titers reaching ~ 0.5 g/L under shake-flask conditions in glucose-based medium (Ledesma-Amaro et al. 2015a).

Parallel strategies focused on improving riboflavin production from plant oils, the preferred carbon source in industrial riboflavin production, by overexpressing genes involved in the glyoxylate cycle (Fig. 3.2), such as the *AgICL1* (encoding isocitrate lyase) (Boeddecker et al. 1997; Maeting et al. 1999) and the *AgMLS1* (encoding malate synthase) (Sugimoto et al. 2009). Additionally, riboflavin excretion into the culture medium was improved by disrupting *AgVMA1*, which encodes the vacuolar ATPase responsible for riboflavin compartmentalization (Förster et al. 1999).

Beyond pathway-specific engineering, emerging evidence points to a critical role for intracellular stress responses in triggering riboflavin overproduction. Multiple studies have shown that environmental cues and stress-activated signaling pathways can modulate riboflavin biosynthesis (Kavitha and Chandra 2009, 2014; Walther and Wendland 2012; Nieland and Stahmann 2013; Silva et al. 2019c; Kato et al. 2021, 2023, 2024). As such, future engineering efforts may benefit from a deeper understanding of these regulatory mechanisms.

Altogether, these multifaceted strategies have positioned *A. gossypii* as a robust microbial platform for sustainable riboflavin production, leveraging metabolic rewiring, carbon source optimization, and regulatory insights derived from decades of research.

3.2.1.2 Flavin Adenine Dinucleotide (FAD), Purine Nucleosides, and Folates

As a natural riboflavin overproducer, *A. gossypii* displays a significantly enhanced metabolic flux through the purine biosynthetic pathway. This not only enables high-level riboflavin production but also supports the synthesis of other valuable purine-derivatives, including FAD (Patel and Chandra 2020), purine nucleosides such as inosine and guanosine (Ledesma-Amaro et al. 2015b; 2016; Buey et al. 2015), and folates (Serrano-Amatriain et al. 2016).

In *A. gossypii*, FAD is synthesized from riboflavin via the sequential action of riboflavin kinase (AgFmn1) and FAD synthetase (AgFad1) (Fig. 3.2). Compared to riboflavin, which is primarily used as a dietary supplement, FAD holds greater pharmaceutical relevance and is therefore more expensive. Gene expression and metabolite profiling studies identified the reaction catalyzed by AgFmn1 as a major bottleneck in FAD biosynthesis. Accordingly, overexpression of *AgFMN1* resulted in a 14-fold increase in FAD levels, with production titers reaching ~ 0.09 g/L under shake-flask conditions (Patel and Chandra 2020). However, to match the titers achieved by other microorganisms via *de novo* synthesis (~ 0.45 g/L), further genetic and bioprocess optimizations are necessary.

Inosine and guanosine are commercially important compounds in the food industry, valued for their umami-enhancing properties and associated health benefits. In the *de novo* purine biosynthesis pathway, inosine monophosphate (IMP) is produced from ribose-5-phosphate through a series of enzymatic reactions. IMP then serves as a precursor for the synthesis of AMP and GMP. These and other nucleotides in this pathway can be hydrolyzed to nucleosides, which may either be secreted or degraded into nucleobases for reutilization via the purine salvage pathway (Fig. 3.2). Through targeted metabolic engineering, particularly the deletion of *AgADE12* (encoding for adenylosuccinate synthase) and *AgPNP1* (encoding for purine nucleoside phosphorylase), the *A. gossypii* metabolic flux was redirected away from AMP synthesis and nucleoside degradation. As a result, inosine production increased by up to 200-fold (~ 2.24 g/L), and guanosine by threefold (~ 0.16 g/L) in shake-flask cultures (Ledesma-Amaro et al. 2015b; 2016). Notably, deletion of the *AgIMD3/4* gene, encoding the IMP dehydrogenase that catalyzes the oxidation of IMP to xanthosine 5′monophosphate (XMP), impaired guanosine production (Buey et al. 2015; Ledesma-Amaro et al. 2016). These results positioned *A. gossypii* as the first eukaryotic microorganism with demonstrated potential for industrial inosine production. Nevertheless, titers remain below those achieved with engineered bacterial strains (6–7 g/L), highlighting the need for further strain and bioprocess development.

Folates (vitamin B9) are essential micronutrients, and folate deficiency has been linked to several serious health conditions. To address this, many countries have mandated food fortification with synthetic folic acid. Since both riboflavin and folate biosynthesis share GTP as precursor, *A. gossypii* with its naturally elevated GTP pool emerged as a promising chassis for folate production (Revuelta et al. 2018). In *A. gossypii*, the folate biosynthetic pathway converts GTP to tetrahydrofolate (THF), the first biologically active form of folate, via eight enzymatic steps (Fig. 3.2). Three key enzymes, encoded by *AgFOL1*, *AgFOL2*, and *AgFOL3*, drive the conversion of GTP to 7,8-dihydrofolate (DHF). Among them, GTP cyclohydrolase I (*AgFOL2*) competes directly with Rib1 (GTP cyclohydrolase II) for the GTP pool. *AgFOL1* encodes a multifunctional enzyme with dihydroneopterin aldolase (DHNA), hydroxymethylpterin pyrophosphokinase (HPPK), and dihydropteroate synthetase (DHPS) activities, while *AgFOL3* encodes DHF synthetase (DHFS). Overexpressing the three *AgFOL* genes in *A. gossypii* led to a 16-fold increase in folate production (Serrano-Amatriain et al. 2016). The final step, reduction of DHF to THF, is catalyzed by DHF reductase (DHFR), encoded by *AgDFR1*. Intracellular folate retention and activity are enhanced by polyglutamylation, a modification catalyzed by the *AgMET7*-encoded folylpoly-γ-glutamate synthetase (FPGS), which strengthens the affinity of folate-dependent enzymes for their cofactors. Deletion of *AgMET7* in *A. gossypii* led to a 5.7-fold increase in folate production, likely by relieving feedback inhibition from folyl-polyglutamates and reducing intracellular folate retention (Serrano-Amatriain et al. 2016).

Building on the findings by Ledesma-Amaro et al. (2015b, 2016), deletion of *AgADE12* redirects the *A. gossypii* metabolic flux toward the GMP branch, ultimately enhancing GTP availability for folate synthesis. Therefore, overexpression of the three *AgFOL* genes and deletion of *AgMET7* were combined with the deletion of *AgADE12*, resulting in a 51-fold improvement in folate production (Serrano-Amatriain et al. 2016). Additional enhancement was achieved by diverting GTP from riboflavin to folate biosynthesis through underexpression of the *AgRIB1* gene. This final engineered strain achieved folate titers of 0.007 g/L in shake-flask cultures (146-fold higher than the unmodified strain) and produced folate profiles enriched in 5-methyl-THF, closely resembling those found in natural folate-rich foods (Serrano-Amatriain et al. 2016). Despite these remarkable results, the best reported to date for microbial folate bioproduction, further strain development and process optimization in bioreactors will be essential to achieve the high titers, yields, and productivities required for a commercially viable and sustainable alternative to chemically synthesized folic acid (Revuelta et al. 2018).

3.2.2 *de novo* Pyrimidine Biosynthesis Pathway

Using a Cre-*loxP*-based targeted gene disruption strategy (Aguiar et al. 2014a), the *A. gossypii de novo* pyrimidine biosynthesis pathway was disrupted at the orotidine-5'-phosphate decarboxylase (AgUra3) level to generate an uracil/uridine auxotroph (Fig. 3.3). This disruption not only produced the expected auxotrophy but also led to orotic acid accumulation (Silva et al. 2022) and enhanced riboflavin production (Silva et al. 2015).

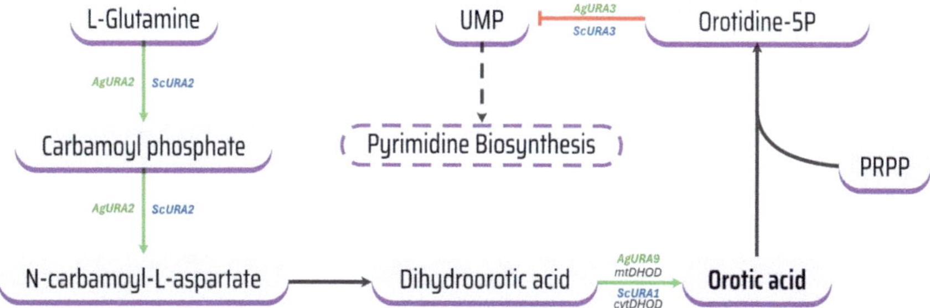

Fig. 3.3 Schematic representation of metabolic engineering strategies targeting the *Ashbya gossypii de novo* pyrimidine biosynthesis pathway for enhanced orotic acid production, with key genetic targets highlighted. Dashed arrows indicate a multistep pathway, green arrows indicate native gene overexpression; red line indicates gene knockout. cytDHOD, cytosolic dihydroorotate dehydrogenase; mtDHOD, mitochondrial dihydroorotate dehydrogenase; orotidine 5-P, orotidine 5-phosphate; PRPP, phosphoribosyl pyrophosphate; UMP, uridine monophosphate

Subsequent studies proposed two mechanisms for the riboflavin overproduction phenotype of this *Agura3* mutant: (i) nutritional stress from uracil/uridine starvation activates riboflavin biosynthesis (Silva et al. 2015); and (ii) increased availability of phosphoribosyl pyrophosphate (PRPP) for riboflavin biosynthesis, due to reduced consumption by the blocked pyrimidine pathway (Silva et al. 2019b).

Regarding orotic acid accumulation, the mitochondrial dihydroorotate dehydrogenase (DHOD) AgUra9 was found to play a key role. Heterologous expression of *AgURA9* in a *S. cerevisiae Scura3 Scura1* double mutant enabled that strain to accumulate and excrete orotic acid, demonstrating that AgUra9 is determinant for this phenotype (Silva et al. 2022).

Given the biotechnological potential of the *A. gossypii Agura3* strain, its native ability to utilize glycerol, and the industrial interest in valorizing crude glycerol, this strain was subjected to further metabolic engineering. Since overexpression of the *AgGUP1*-encoded glycerol transporter was previously shown to improve glycerol utilization and hyperosmotic stress tolerance in glycerol-containing media (Silva et al. 2014), *AgGUP1* was placed under the strong, constitutive P_{AgGPD} promoter in the *Agura3* background to boost glycerol consumption and osmotic stress tolerance (Silva et al. 2022). In this engineered chassis, combined overexpression of *AgURA2* and *AgURA9* further increased orotic acid production from crude glycerol, with titers reaching ~ 4.5 g/L in shake-flask culture under optimized conditions (Silva et al. 2022).

3.2.3 Xylose Assimilation Pathway

Driven by the principles of the bioeconomy and circular economy, the use of renewable feedstocks such as lignocellulosic biomass has become a central strategy in the development of sustainable bioprocesses. In this context, the metabolic potential of *A. gossypii* to assimilate C5 sugars, particularly *D*-xylose, given its predominance in lignocellulosic hydrolysates, has attracted considerable interest (Ribeiro et al. 2012; Díaz-Fernández et al. 2017; Montero-Bullón et al. 2025a).

Initial studies by Ribeiro et al. (2012) demonstrated that although *A. gossypii* is unable to grow on *D*-xylose as the sole carbon source, it is capable of converting it into xylitol. This finding indicated the presence of functional pentose transporters and suggested that the native *A. gossypii* xylose assimilation pathway is cryptically expressed. Accordingly, subsequent metabolic engineering strategies focused on activating this pathway by overexpressing the genes encoding its key enzymes (Fig. 3.4): (i) xylose reductase (*AgGRE3*), (ii) xylitol dehydrogenase (*AgXYL2*), and (iii) xylulose kinase (*AgXKS1*). Simultaneous overexpression of these genes enabled the efficient conversion of *D*-xylose into xylulose-5-phosphate, which enters the non-oxidative branch of the pentose phosphate pathway, thereby supporting both cellular growth and biosynthesis of value-added metabolites (Díaz-Fernández et al. 2017). Furthermore, a slight improvement in xylose utilization in the presence of glucose was achieved by overexpressing the Afl205Cp hexose transporter carrying the N355V amino acid substitution in this strain, which was identified as playing a key role in xylose uptake (Díaz-Fernández et al. 2020).

To further enhance the metabolic capabilities of the xylose-consuming engineered *A. gossypii* strain, particularly for the production of acetyl-CoA-derived compounds such as fatty acids and terpenoids, a heterologous phosphoketolase pathway was introduced to redirect the carbon flux from xylulose-5-phosphate directly toward acetyl-CoA biosynthesis. For this purpose, the coding sequences of *xkpA* from *Aspergillus nidulans* (encoding a phosphoketolase) and *pta* from *Bacillus subtilis* (encoding a phosphotransacetylase) were integrated into its genome under the control of the strong constitutive promoter P_{GPD1}, resulting in the engineered strain *A. gossypii* A729 (Díaz-Fernández et al. 2017).

Together, these metabolic engineering strategies positioned this *A. gossypii* strain as a promising chassis for sustainable bioproduction from xylose-rich lignocellulosic feedstocks, leveraging its enhanced xylose metabolism and acetyl-CoA-generating capabilities to produce a broad range of high-value biochemicals, including microbial lipids and terpenoids (Díaz-Fernández et al. 2019; Muñoz-Fernández et al. 2022, 2024; Montero-Bullón et al. 2025a).

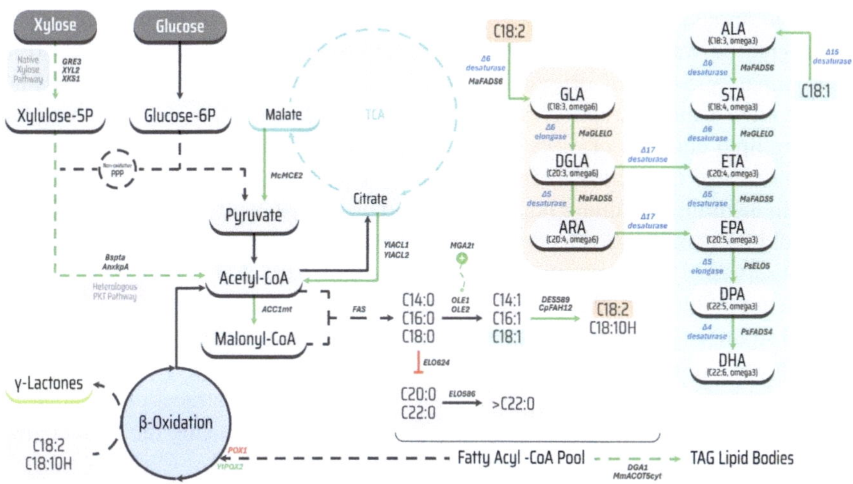

Fig. 3.4 Schematic representation of the interconnected xylose assimilation, fatty acid biosynthesis, elongation/desaturation, and β-oxidation pathways in *Ashbya gossypii*, highlighting shared precursors, branch points, and key metabolic engineering targets for enhanced production of microbial lipids, omega-3 and omega-6 polyunsaturated fatty acids (PUFAS), and γ-lactones. Dashed arrows indicate a multistep pathway, green arrows/gene names indicate gene overexpression; green broken line indicates transcription activation of *AgOLE1* by the truncated transcription factor AgMga2t; red lines/gene names indicate gene knockout. ALA, alpha-linolenic acid; ARA, arachidonic acid; C18:1, oleic acid; C18:2, linoleic acid; C18:10H, ricinoleic acid; EPA, eicosapentaenoic acid; ETA, eicosatetraenoic acid; DGLA, di-homo-gamma-linolenic acid; DHA, docosahexaenoic acid; DPA, docosapentaenoic acid; GLA, gamma-linolenic acid; glucose-6P, glucose-6 phosphate; PKT, phosphoketolase; PPP, pentose phosphate pathway; STA, stearidonic acid; TAG, triacylglycerol; xylulose-5P, xylulose-5 phosphate

3.2.4 Fatty Acid Biosynthesis, Elongation/Desaturation, and β-oxidation Pathways

3.2.4.1 Microbial Lipids

A. gossypii was first reported to accumulate moderate amounts of lipid droplets within its hyphae as early as 1994 (Stahmann et al. 1994), but its potential as a microbial oil production chassis was not actively investigated until 2014 (Ledesma-Amaro et al. 2014a; 2014b). Since then, various metabolic engineering strategies targeting fatty acid (FA) biosynthesis and degradation pathways (Fig. 3.4) have been employed to convert *A. gossypii* into an oleaginous organism, capable of accumulating more than 20–25% of its dry weight as lipids. Initial modifications included the introduction of ATP-citrate lyase (ACL) activity to enhance cytosolic acetyl-CoA availability for lipid biosynthesis, and the disruption of lipid degradation via the β-oxidation pathway through deletion of the unique acyl-CoA oxidase-encoding gene (*AgPOX1*) and/or *AgFOX2*, which encodes a multifunctional β-oxidation enzyme (Ledesma-Amaro et al. 2014a).

Simultaneous overexpression of the heterologous *YlACL1* and *YlACL2* genes from *Yarrowia lipolytica* resulted in a ~ 30% increase in intracellular lipid accumulation when glucose was used as the sole carbon source, raising the lipid content from approximately 5–7% of dry cell weight (DCW) (Ledesma-Amaro et al. 2014a). On the other hand, disruption of the β-oxidation pathway led to a ~ 45% increase, yielding lipid contents of ~ 9% of DCW (Ledesma-Amaro et al. 2014a). Notably, deletion of *AgPOX1* enabled intracellular lipid accumulation of up to ~ 70% of DCW in medium containing both glucose and oleic acid, well above the established threshold for classification as an oleaginous microorganism and far exceeding the ~ 20% observed in the unmodified strain (Ledesma-Amaro et al. 2014a).

Deletion of the *AgPOX1* gene in the previously described *A. gossypii* A729 background strain, engineered to channel xylose flux toward microbial lipid biosynthesis through overexpression of the endogenous xylose-utilizing pathway and an heterologous phosphoketolase pathway, led to a 30–45% increase in intracellular lipid accumulation when xylose was used as the sole carbon source, raising the lipid content from 7–8% to 10–13% of DCW (Díaz-Fernández et al. 2017).

In a subsequent study, expression of a C-terminal truncation of AgMga2/Spt23 (AgMga2-ΔC-term), a regulator of the main Δ9 desaturase gene (*AgOLE1*), along with overexpression of the heterologous *MmACOT5* gene from *Mus musculus*, which encodes a fatty acyl-CoA thioesterase involved in peroxisomal FA degradation, enabled lipid accumulation of up to ~ 25% of DCW in molasses-based medium (~ 75% more than the unmodified strain) (Lozano-Martínez et al. 2017).

Comparable improvements in lipid accumulation using xylose as the sole carbon source were achieved by combining, in the previously described *A. gossypii* A729 background strain, the expression of a C-terminal truncation of AgMga2/Spt23 with substitution of the native acetyl-CoA carboxylase AgAcc1 by a feedback resistant mutant (AgAcc1^{S659A,S1157A}), and overexpression of an additional copy of the native

DGA1 gene, which encodes a diacylglycerol acyltransferase involved in triacylglycerol biosynthesis (Díaz-Fernández et al. 2019). When cultivated in detoxified corn cob hydrolysate plus molasses or crude glycerol, this new strain with a deregulated lipid anabolism (*A. gossypii* A877) accumulated up to ~ 40% of DCW in lipid content, reaching a lipid titer of 4 g/L under shake-flask conditions (Díaz-Fernández et al. 2019).

In a follow-up study, several of the above-mentioned strategies were combined in a newly constructed background strain, *A. gossypii* A1141, which shares the same metabolic phenotype as A729 but carries fewer genomic scars (Montero-Bullón et al. 2025a). Engineering strain A1141 with the following modifications led to a substantial increase in lipid content, reaching ~ 50% of DCW using xylose as sole carbon source: (i) overexpression of the heterologous *YlACL1* and *YlACL2* genes; (ii) overexpression of the deregulated AgAcc1^{S659A,S1157A} mutant; (iii) truncation of the C-terminal domain of Mga2; and (iv) overexpression of an additional copy of the endogenous *DGA1* gene (Montero-Bullón et al. 2025a).

Further improvement (~ 75% relative to A1141) was achieved by deleting the *AgPOX1* gene in this newly engineered A1576 strain and replacing it with an overexpression cassette for the heterologous *McMCE2* gene from *Mucor circinelloides*, which encodes a cytosolic NADP$^+$-dependent malic enzyme (NADP-ME) involved in NADPH regeneration, a rate-limiting cofactor in lipid biosynthesis (Montero-Bullón et al. 2025a). This engineering strategy enabled the final strain, *A. gossypii* A1623, to accumulate up to ~ 60% of DCW using xylose as sole carbon source (Montero-Bullón et al. 2025a).

The characterization and engineering of both native and heterologous elongation and desaturation systems have also enabled the customization of the *A. gossypii* FA profile (Ledesma-Amaro et al. 2014b; 2018; Lozano-Martínez et al. 2017). The FA composition in *A. gossypii* is determined by its fatty acid synthase (FAS) complex, along with downstream elongase and desaturase enzymes. The *A. gossypii* FAS complex synthesizes saturated FAs with 16 and 18 carbon atoms (C16:0 and C18:0), which can be further elongated by two elongases, AgElo624 (AFR624Wp) and AgElo586 (AFR586Wp), producing FAs of up to C22:0 and C26:0, respectively. Monounsaturated derivatives of these FAs are synthesized by two native Δ9 desaturases, AAL078Wp (AgOle2) and AAR153Cp (AgOle1), homologs of the *S. cerevisiae* ScOle1. Linoleic acid (C18:2) is also naturally synthesized from oleic acid (C18:1) by a native Δ12 desaturase, AgDes589 (AFR589Cp), which is absent in *S. cerevisiae* (Ledesma-Amaro et al. 2014b) (Fig. 3.4).

Oleic acid is the predominant FA, accounting for over 50% of the total FA pool. It serves as key precursor for the biosynthesis of value-added fatty acids, including linoleic acid (Fig. 3.4), the direct precursor of omega-3 and omega-6 polyunsaturated fatty acids (PUFAs) (Ledesma-Amaro et al. 2018) and an indirect precursor for δ-lactones (Silva et al. 2021), as well as ricinoleic acid (Lozano-Martínez et al. 2017), a precursor of γ-lactones (Silva et al. 2019a; 2021). Accordingly, the most significant modifications in the highly malleable *A. gossypii* FA profile have focused on enhancing linoleic acid accumulation. For example, overexpression of the Δ12 desaturase (*AgDES589*) resulted in a fivefold increase in linoleic acid levels (up to

16% of total FAs) compared to the unmodified strain. Moreover, co-overexpression of both native Δ9 desaturase genes (*AAL078W* and *AAR153C*) along with the heterologous *CpFAH12* gene from *Claviceps purpurea*, which encodes a bifucntional oleate 12-hydroxylase/desaturase capable of inserting either a hydroxyl group or a double bond at the C12 position of oleic acid, led to a 15-fold increase in linoleic acid levels compared to the unmodified strain, reaching up to 20% of total FA (Lozano-Martínez et al. 2017). In this strain, ricinoleic acid was also produced but in modest amounts, suggesting that the primary contribution of CpFah12 in *A. gossypii* is related to its Δ12 desaturase activity (Lozano-Martínez et al. 2017).

Building on the enhanced linoleic acid-producing phenotype of the previously described *A. gossypii* strain overexpressing the native Δ12 desaturase, this strain was further explored as chassis for the production of PUFAS (Ledesma-Amaro et al. 2018). To enable PUFA biosynthesis, three metabolic engineering strategies were implemented to equip the strain with the enzymatic machinery required for the downstream elongation and desaturation of linoleic acid into long-chain omega-3/6 PUFAs.

Co-overexpression of the heterologous arachidonic acid (ARA; C20:4) biosynthetic pathway from *Mortierella alpina*, comprising the Δ6 desaturase (*MaFADS6*), Δ6 elongase (*MaGLELO*), and Δ5 desaturase (*MaFADS5*), along with the endogenous Δ12 desaturase, enabled the production of omega-6 PUFAs as ~ 24% of total FAs, with ARA constituting up to 5% of the total FA content (Ledesma-Amaro et al. 2018). Further engineering by co-overexpressing a Δ17 desaturase gene from *Phytophthora infestans* enabled the conversion of omega-6 to omega-3 PUFAs, leading to the production of eicosapentanoic acid (EPA; C20:5), alpha-linolenic acid (ALA; C18:3) and eicosatetraenoic acid (ETA; C20:4), with a total omega-3 PUFA content reaching ~ 5% of total FAs (Ledesma-Amaro et al. 2018). Additional overexpression of an heterologous omega-3 docosahexaenoic acid (DHA; C22:6) biosynthetic pathway, comprising a Δ12/Δ15 desaturase from *Fusarium moniliforme*, and the Δ5 elongase (*PsELO5*) and Δ4 desaturase (*PsFADS4*) from *Pavlova salina*, further increased EPA levels (to ~ 4% of total FAs) and enabled the production of docosapentaenoic acid (DPA; C22:5) and DHA, with the total PUFA content reaching 20% (Ledesma-Amaro et al. 2018). These results highlighted the key role of Δ17 desaturase for the conversion of omega-6 to omega-3 PUFAs by *A. gossypii*, including the complete conversion of ARA to EPA, and underscored the potential of *A. gossypii* as a microbial platform for the production of high-value functional lipids such as ARA, EPA, and DHA (Ledesma-Amaro et al. 2018).

3.2.4.2 Lactones

Characterization of the volatile organic profile of various *A. gossypii* strains cultivated in a yeast extract–peptone–glucose medium revealed a remarkable native capacity for the biosynthesis of γ-lactones, particularly γ-decalactone (Silva et al. 2019a).

Given their distinctive fruity aromas and flavor notes, γ-lactones are of considerable commercial interest as food additives and fragrance ingredients, with additional potential in various industrial applications (Silva et al. 2021).

To understand and enhance the *de novo* biosynthesis of γ-lactones from glucose (Fig. 3.4), metabolic engineering strategies were developed targeting key steps in the FA biosynthesis and *β*-oxidation pathways. Notably, overexpression of *AgDES589*, which encodes the native Δ12 desaturase that converts oleic acid into linoleic acid, led to a ~ sixfold increase in specific γ-lactone production (~ 0.04 g/L under shake-flask conditions) (Silva et al. 2019a). Interestingly, this increase was not uniform across all γ-lactones. In parallel, deletion of *AgELO624*, encoding a native elongase responsible for the synthesis of very-long-chain FAs (C20–C26), resulted in a ~ threefold increase in lactone production, specifically in γ-decalactone, suggesting that very-long-chain FAs are suboptimal precursors for γ-lactone biosynthesis (Silva et al. 2019a). Combination of both modifications led to a fourfold increase in total γ-lactone production compared to the unmodified strain, mainly boosting γ-decalactone and γ-octalactone levels (Silva et al. 2019a). These findings demonstrated that redirecting the FA flux at the C18 level, particularly from oleic acid, the major FA naturally accumulated by *A. gossypii*, significantly influences γ-lactone biosynthesis, thus supporting its role as the main precursor for γ-lactone production in this fungus (Silva et al. 2019a; 2021).

Further pathway optimization was achieved by replacing *AgPOX1*, which encodes the sole acyl-CoA oxidase in *A. gossypii*, with an overexpression cassette for the *YlPOX2* from *Yarrowia lipolytica*. YlPox2 preferentially oxidizes long-chain acyl-CoAs (C11–C18), leading to a shift in the γ-lactone profile, with γ-decalactone accounting for over 99% of the total γ-lactone content (Silva et al. 2019a).

Altogether, these combinatorial metabolic engineering strategies not only elucidated the biosynthetic pathway but also significantly enhanced the production of γ-decalactone, underscoring the potential of *A. gossypii* as a promising microbial platform for the sustainable production of these and other (Birk et al. 2019; Silva et al. 2021) high-value lactones.

3.2.5 *Mevalonate Pathway and Terpene Biosynthesis*

Terpenes (also known as terpenoids or isoprenoids) are a large and structurally diverse class of naturally occurring compounds composed of C5 isoprene units, specifically isopentenyl diphosphate (IPP) and dimethylallyl diphosphate (DMAPP). In *A. gossypii*, these essential building blocks are synthesized from acetyl-CoA via the mevalonate (MVA) pathway (Fig. 3.5). This pathway begins with the condensation of acetyl-CoA into acetoacetyl-CoA, catalyzed by AgErg10, followed by conversion to 3-hydroxy-3-methylglutaryl-CoA (HMG-CoA) via AgErg13. The rate-limiting enzyme HMG-CoA reductase (AgHmg1) then converts HMG-CoA to mevalonate. Subsequent enzymatic steps involving AgErg12, AgErg8, and AgErg19 convert mevalonate into IPP, which can then be isomerized into DMAPP by AgIdi1. These C5 precursors can be condensed by AgErg20 and AgBts1 to generate longer-chain prenyl diphosphates, including geranyl diphosphate (GPP, C10), farnesyl diphosphate (FPP, C15), and geranylgeranyl diphosphate (GGPP, C20), which serve as immediate precursors of monoterpenes, sesquiterpenes, and diterpenes, respectively (Muñoz-Fernández et al. 2022; 2024).

Given the natural ability of certain *A. gossypii* strains to produce acyclic monoterpenes (Semenova et al. 2022), the growing market demand for monoterpenes across various industrial sectors, and the limitations associated with their extraction from natural sources, *A. gossypii* has emerged as a promising microbial chassis for the sustainable bioproduction of monoterpenes (Muñoz-Fernández et al. 2022; 2024).

Engineered *A. gossypii* strains have been developed to produce limonene and sabinene from xylose, using the previously characterized xylose-utilizing strain *A. gossypii* A665 as genetic background. To enable limonene production, a heterologous overexpression cassette encoding a truncated form of the *Citrus limon* limonene synthase ClLs1, lacking its plastid targeting signal (referred to as tLS), was integrated into the genome of *A. gossypii* A665. This was combined with the overexpression of different isoforms of the native AgHmg1, including the full-length version and two truncated variants. Among these, the full-length AgHmg1 isoform yielded the highest limonene production (~ 0.8 mg/L), outperforming the truncated versions (Muñoz-Fernández et al. 2022).

To further enhance production, these modifications were combined with the overexpression of a heterologous phosphoketolase pathway (described above) to increase

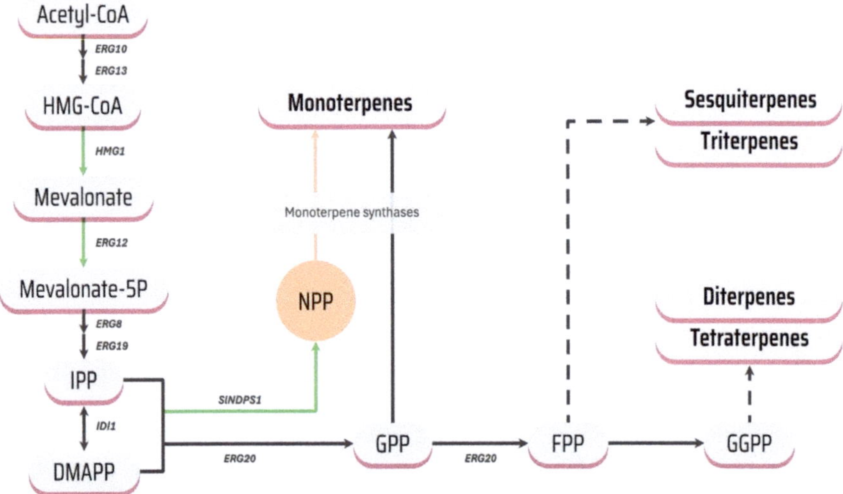

Fig. 3.5 Schematic representation of metabolic engineering strategies targeting the *Ashbya gossypii* mevalonate and terpene biosynthesis pathways for enhanced monoterpene production, with key genetic targets highlighted. Green arrows indicate gene overexpression. DMAPP, dimethylallyl diphosphate; FPP farnesyl diphosphate; GGPP, geranylgeranyl diphosphate; GPP geranyl diphosphate; HMG-CoA, 3-hydroxy-3-methylglutaryl coenzyme A; NPP, neryl diphosphate; IPP, isopentenyl diphosphate, 5P, 5 phosphate. The NPP orthogonal pathway is colored in orange

the cytosolic acetyl-CoA pool, along with a truncated version of the *Solanum lycopersicum SlNDPS1* gene (also lacking the plastid targeting signal), encoding a neryl diphosphate (NPP) synthase that generates NPP, a the cis-isomer of GPP. This combination led to an ~ 42-fold increase in limonene titers (Muñoz-Fernández et al. 2022).

Additional improvement was achieved by overexpressing the native *AgERG12* gene in this engineered background. Altogether, this orthogonal metabolic engineering strategy enabled the final strain, *A. gossypii* A1308, to achieve limonene titers of up to 0.3 g/L from xylose under shake-flask conditions, representing a ~ 420-fold improvement over the initial limonene-producing strains (Muñoz-Fernández et al. 2022; 2024). Similarly, by replacing tLS with a truncated version of the *Salvia pomifera* sabinene synthase lacking its plastid targeting signal (referred to as tSS), sabinene titers of up to 0.5 g/L were obtained from xylose under the same conditions (Muñoz-Fernández et al. 2024). Furthermore, when cultivated in mixed formulations of corn-cob lignocellulosic hydrolysates and either sugarcane or beet molasses, limonene and sabinene titers increased to 0.4 g/L and 0.7 g/L, respectively, the latter representing one of the highest reported titers for sabinene production in flask cultures (Muñoz-Fernández et al. 2024).

Altogether, these results provided proof of principle that *A. gossypii* can be harnessed for the microbial production of monoterpenes such as limonene and sabinene from xylose-based carbon sources. Moreover, the ability of these engineered strains to utilize agro-industrial wastes, such as xylose-rich hydrolysates, positions *A. gossypii* as a promising chassis for sustainable terpene bioproduction.

3.2.6 Ehrlich Pathway

Analysis of the *A. gossypii* aroma alcohol profile has shown that *A. gossypii* generates significant amounts of isoamyl alcohol and 2-phenylethanol, two fusel alcohols with significant biotechnological interest due to their applications as flavoring agents in the food, beverage, and fragrance industries (Ravasio et al. 2014). These compounds are synthesized via the Ehrlich pathway (Fig. 3.6), a catabolic route that converts amino acids into fusel alcohols through three sequential enzymatic reactions: transamination, decarboxylation, and alcohol dehydrogenase reaction. Functional analysis of the *AgARO* genes involved in this pathway demonstrated that deletion of *AgARO8b, AgARO10,* and *AgARO80* strongly impaired aroma compound production, especially 2-phenylethanol, thereby highlighting their essential role in fusel alcohol biosynthesis (Ravasio et al. 2014). In contrast, overexpression of the *AgARO80* gene, a transcriptional regulator of several *AgARO* genes in the Ehrlich pathway (Fig. 3.6), resulted in a 50% increase in overall volatile organic compound production, with isoamyl alcohol and isobutanol levels rising approximately 2.5-fold (Ravasio et al. 2014). Notably, 2-phenylethanol production remained unchanged, likely due to phenylalanine limitation in the culture medium (Ravasio et al. 2014). Supporting this hypothesis, the 2-phenylethanol titer achieved in this study using

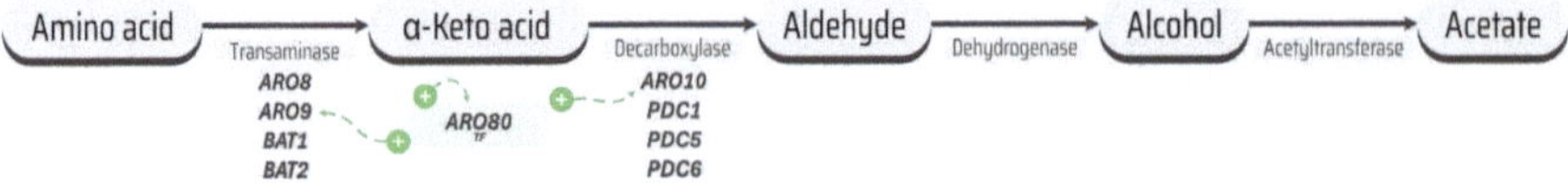

Fig. 3.6 Schematic representation of the *Ashbya gossypii* Ehrlich pathway, with a key genetic target for enhanced production of aroma alcohols highlighted in green

a medium composed of yeast extract, casein peptone, and glucose was markedly lower than titers subsequently reported (~ 0.5 g/L) in industrial *A. gossypii* riboflavin fermentation processes using complex media based on soybean oil and soybean meal (Birk et al. 2019). This contrast underscores the potential for substantial improvement in 2-phenylethanol production through a combined approach of strain engineering and bioprocess optimization.

3.2.7 Ganglioside Biosynthesis and Accessory Pathways

Gangliosides are essential glycosphingolipids involved in key physiological processes, including neuronal differentiation and apoptosis. Traditionally sourced from animal tissues, their production faces ethical, economic, and scalability challenges. Although chemoenzymatic approaches have been developed as alternatives, they lack the flexibility and industrial scalability of microbial systems. Among gangliosides, GM3 and GD3 are of particular interest due to their therapeutic potential and their role as precursors to more complex gangliosides (Montero-Bullón et al. 2025b).

Ganglioside biosynthesis is a tightly regulated, multi-step process that occurs primarily in the Golgi apparatus and which involves a cascade of glycosyltransferase-mediated reactions. The pathway begins with the synthesis of ceramide, composed of a sphingoid base linked to a fatty acid, followed by the sequential addition of glucose and galactose to form glucosylceramide (GlcCer) and lactosylceramide (LacCer), respectively. LacCer then serves a key branching point for ganglioside synthesis (Fig. 3.7). While *A. gossypii* naturally synthesizes GlcCer, it lacks the capacity to convert GlcCer into LacCer (Montero-Bullón et al. 2025b; Gomes et al. 2014). In addition, the synthesis of sialylated gangliosides such as GM3 and GD3 requires N-acetylneuraminic acid (NANA) in its activated form, CMP-NANA, which is generated via the concerted action of three enzymes: UDP-N-acetylglucosamine 2-epimerase/N-acetylmannosamine kinase (GNE), N-acetylneuraminic acid synthase (NANS), and CMP-N-acetylneuraminic acid synthetase (CMAS). Once available, CMP-NANA enables the formation of GM3 by GM3 synthase (ST3Gal5) through sialylation of LacCer, and subsequently GD3 via GD3 synthase (ST8Sia1) through the addition of a second sialic acid (Montero-Bullón et al. 2025b).

To enable *de novo* production of GM3 and GD3 in *A. gossypii*, a modular metabolic engineering strategy was employed, involving the genomic integration

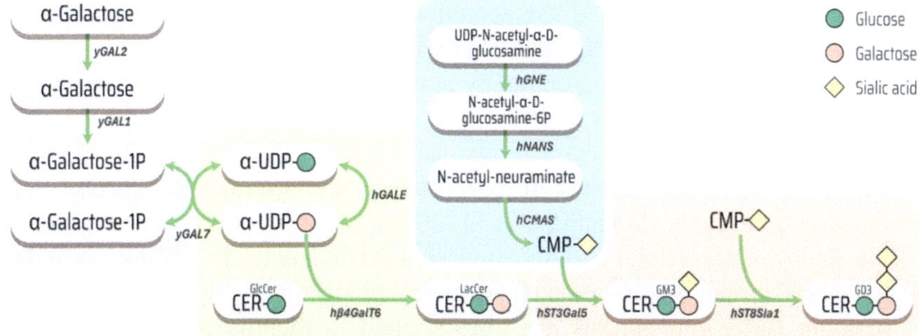

Fig. 3.7 Schematic representation of the pathways that were introduced into the *Ashbya gossypii* genome to achieve GD3 ganglioside bioproduction, color-coded as galactose metabolism (gray), LacCer synthesis (green), sialic acid biosynthesis (blue), and ganglioside synthesis (orange). Metabolites are symbolized by shapes: circles for sugars (green, glucose; orange, galactose) and diamonds for sialic acid. CMP-sialic acid, activated form of sialic acid; GlcCer, glucosylceramide; LacCer, lactosylceramide; UDP, uridine diphosphate

of codon-optimized heterologous gene expression cassettes, grouped into four functional modules (Fig. 3.7): (i) the NANA module (containing the human enzymes hGNE, hNANS, and hCMAS) to enable the synthesis of CMP-NANA, (ii) the hGAL module (containing the human enzymes UDP-galactose-4-epimerase hGALE and beta-1,4-galactosyltransferase 6 hß4GalT6) for conversion of GlcCer to LacCer, (iii) the yGAL module (containing the *S. cerevisiae* galactose permease ScGal2, galactokinase ScGal1, and galactose-1-phosphate uridyl transferase ScGal7) to activate the Leloir pathway and enhance the UDP-galactose pool for LacCer synthesis, and (iv) the GX3 module (containing the human sialyl transferases hST3Gal5 and hST8Sia1) for sequential sialylation of LacCer to GM3 and GM3 to GD3. Each module addressed a specific metabolic bottleneck, enabling full reconstruction of the ganglioside biosynthetic pathway in *A. gossypii* and microbial production of GM3 and GD3 at mg/L and µg/L levels, respectively (Montero-Bullón et al. 2025b).

This work established *A. gossypii* as a novel chassis for microbial ganglioside production, laying the groundwork for microbial production of complex gangliosides for therapeutic, biotechnological, and research purposes.

3.2.8 Recombinant Protein Secretion

A. gossypii naturally exhibits high secretory potential due to its filamentous morphology and extensive hyphal network, which provides a large surface area for protein export. However, the secretion of heterologous proteins remains constrained by metabolic and cellular bottlenecks (Aguiar et al. 2015; 2017).

Early studies demonstrated the secretion of functional heterologous proteins such as *Trichoderma reesei* cellulases (EGI and CBHI) and *Aspergillus niger β*-galactosidase, with expression driven by native (P_{AgTEF}, P_{AgGPD}) and heterologous (P_{ScPGK1}, P_{ScADH1}) promoters (Ribeiro et al. 2010; Magalhães et al. 2014). Among these, P_{AgTEF} yielded higher extracellular activity (Magalhães et al. 2014). Medium composition, particularly the use of glycerol as a carbon source, significantly influenced secretion levels (Magalhães et al. 2014). Despite detectable expression of CBHI, its low extracellular activity suggested inefficiencies in folding, trafficking, or post-translational modification (Ribeiro et al. 2010).

Random EMS mutagenesis has been applied to identify *A. gossypii* mutants with enhanced protein secretion profiles, including increased release of endogenous hydrolases such as *β*-glucosidase and *α*-amylase (Ribeiro et al. 2013). In parallel, targeted deletion of genes potentially involved in cell wall integrity (including *ScGAS1* homologs) was tested to improve protein export, though often with trade-offs in growth and viability (Ribeiro et al. 2013).

Importantly, glycosylation plays a central role in protein folding, trafficking, stability, and immunogenicity. *A. gossypii* possesses a functional *N*-glycosylation system, producing *N*-glycan profiles that suggest a predominance of high-mannose-type structures, similar to *S. cerevisiae*, but with possible strain-specific variations in glycan branching and trimming (Aguiar et al. 2013). These features can impact the heterologous production of therapeutics that require human-like glycosylation. As such, glycoengineering strategies such as introducing heterologous glycosidases, mannosidases, or glycosyltransferases may be required to tailor the glycosylation profile for specific applications.

Secretome analyses revealed that only a small portion of the *A. gossypii* proteome is secreted under standard conditions, typically favoring proteins with moderate molecular weights and acidic isoelectric points. Transcriptomic analyses under secretion stress (e.g., DTT treatment) have revealed a limited unfolded protein response (UPR) in *A. gossypii*. Canonical regulators like *AgIRE1*, *AgHAC1*, *AgKAR2*, and *AgPDI1* are present but do not show robust induction, suggesting a noncanonical or attenuated UPR. This may restrict the fungus's capacity to handle large amounts of misfolded protein, thus representing a potential engineering target for improving recombinant protein yield (Aguiar et al. 2014c).

Taken together, these findings indicate that improving recombinant protein secretion in *A. gossypii* requires a multifaceted approach. Effective strategies include the selection of strong promoters, secretion-competent signal peptides, host strain engineering, glycosylation pathway modulation, and potential enhancement of stress response pathways (Aguiar et al. 2015; 2017). Together with its growing molecular toolbox, these efforts are gradually positioning *A. gossypii* as a promising chassis for recombinant protein secretion. However, significant improvements are still needed to achieve secretion levels comparable to those of already established protein production hosts.

References

Aguiar TQ, Maaheimo H, Heiskanen A, Wiebe MG, Penttilä M, Domingues L (2013) Characterization of the *Ashbya gossypii* secreted *N*-glycome and genomic insights into its *N*-glycosylation pathway. Carbohydr Res 381:19–27. https://doi.org/10.1016/j.carres.2013.08.015

Aguiar TQ, Dinis C, Domingues L (2014a) Cre-*lox*P-based system for removal and reuse of selection markers in *Ashbya gossypii* targeted engineering. Fungal Genet Biol 68:1–8. https://doi.org/10.1016/j.fgb.2014.04.009

Aguiar TQ, Dinis C, Magalhães F, Oliveira C, Wiebe MG, Penttilä M, Domingues L (2014b) Molecular and functional characterization of an invertase secreted by *Ashbya gossypii*. Mol Biotechnol 56:524–534. https://doi.org/10.1007/s12033-013-9726-9

Aguiar TQ, Ribeiro O, Arvas M, Wiebe MG, Penttilä M, Domingues L (2014c) Investigation of protein secretion and secretion stress in *Ashbya gossypii*. BMC Genomics 15:1137. https://doi.org/10.1186/1471-2164-15-1137

Aguiar TQ, Silva R, Domingues L (2015) *Ashbya gossypii* beyond industrial riboflavin production: a historical perspective and emerging biotechnological applications. Biotechnol Adv 33:1774–1786. https://doi.org/10.1016/j.biotechadv.2015.10.001

Aguiar TQ, Silva R, Domingues L (2017) New biotechnological applications for *Ashbya gossypii*: challenges and perspectives. Bioengineered 8:309–315. https://doi.org/10.1080/21655979.2016.1234543

Alberti-Segui C, Dietrich F, Altmann-Jöhl R, Hoepfner D, Phillipsen P (2001) Cytoplasmic dynein is required to oppose the force that moves nuclei towards the hyphal tip in the filamentous ascomycete *Ashbya gossypii*. J Cell Sci 114:975–986. https://doi.org/10.1242/jcs.114.5.975

Althöfer H, Revuelta JL (2003) Genetic strain optimization for improving the production of riboflavin. WO Patent 2003048367 A1, 12 Jun 2003. https://patents.google.com/patent/WO2003048367A1/en

Althöfer H, Seulberg H, Zelder O, Revuelta JL (1999) Genetic method for producing riboflavin. WO Patent 1999061623 A2, 2 Dec 1999. https://patents.google.com/patent/WO1999061623A2/en

Altmann-Jöhl R, Philippsen P (1996) *AgTHR4*, a new selection marker for transformation of the filamentous fungus *Ashbya gossypii*, maps in a four-gene cluster that is conserved between *A. gossypii* and *Saccharomyces cerevisiae*. Mol Gen Genet 250:69–80. https://doi.org/10.1007/BF02191826

Ayad-Durieux Y, Knechtle P, Goff S, Dietrich F, Philippsen P (2000) A PAK-like protein kinase is required for maturation of young hyphae and septation in the filamentous ascomycete *Ashbya gossypii*. J Cell Sci 113:4563–4575. https://doi.org/10.1242/jcs.113.24.4563

Bauer Y, Knechtle P, Wendland J, Helfer H, Philippsen P (2004) A Ras-like GTPase is involved in hyphal growth guidance in the filamentous fungus *Ashbya gossypii*. Mol Biol Cell 15:4622–4632. https://doi.org/10.1091/mbc.e04-02-0104

Birk F, Fraatz MA, Esch P, Heiles S, Pelzer R, Zorn H (2019) Industrial riboflavin fermentation broths represent a diverse source of natural saturated and unsaturated lactones. J Agric Food Chem 67:13460–13469. https://doi.org/10.1021/acs.jafc.9b01154

Buey RM, Ledesma-Amaro R, Balsera M, de Pereda JM, Revuelta JL (2015) Increased riboflavin production by manipulation of inosine 5'-monophosphate dehydrogenase in *Ashbya gossypii*. Appl Microbiol Biotechnol 99:9577–9589

DeMay BS, Meseroll RA, Occhipinti P, Gladfelter AS (2010) Cellular requirements for the small molecule forchlorfenuron to stabilize the septin cytoskeleton. Cytoskeleton (Hoboken) 67:383–399. https://doi.org/10.1002/cm.20452

Dessipri E, Smith J, Walch S, Srinivasan J (2022) Riboflavin from *Ashbya gossypii*. In: 92nd JECFA—Chemical and Technical Assessment (CTA). FAO, Rome, p 13. https://openknowledge.fao.org/server/api/core/bitstreams/70910278-fc3b-460e-9197-a131d5477619/content

Díaz-Fernández D, Lozano-Martínez P, Buey RM, Revuelta JL, Jiménez A (2017) Utilization of xylose by engineered strains of *Ashbya gossypii* for the production of microbial oils. Biotechnol Biofuels 10:3. https://doi.org/10.1186/s13068-016-0685-9

Díaz-Fernández D, Aguiar TQ, Martín VI, Romaní A, Silva R, Domingues L, Revuelta JL (2019) Microbial lipids from industrial wastes using xylose-utilizing *Ashbya gossypii* strains. Bioresour Technol 293:122054. https://doi.org/10.1016/j.biortech.2019.122054

Díaz-Fernández D, Muñoz-Fernández G, Martín VI, Revuelta JL, Jiménez A (2020) Sugar transport for enhanced xylose utilization in Ashbya gossypii. J Ind Microbiol Biotechnol 47:1173–1179. https://doi.org/10.1007/s10295-020-02320-5

Dietrich FS, Voegelli S, Kuo S, Philippsen P (2013) Genomes of *Ashbya* fungi isolated from insects reveal four mating-type loci, numerous translocations, lack of transposons, and distinct gene duplications. G3 (Bethesda) 3:1225–1239. https://doi.org/10.1534/g3.112.002881

Dünkler A, Wendland J (2007) Use of *MET3* promoters for regulated gene expression in *Ashbya gossypii*. Curr Genet 52:1–10. https://doi.org/10.1007/s00294-007-0134-1

Dünkler A, Jorde S, Wendland J (2008) An *Ashbya gossypii cts2* mutant deficient in a sporulation-specific chitinase can be complemented by *Candida albicans CHT4*. Microbiol Res 163:701–710. https://doi.org/10.1016/j.micres.2008.08.005

Finlayson MR, Helfer-Hungerbühler AK, Philippsen P (2011) Regulation of exit from mitosis in multinucleate *Ashbya gossypii* cells relies on a minimal network of genes. Mol Biol Cell 22:3081–3093. https://doi.org/10.1091/mbc.e10-12-1006

Föster C, Santos M, Ruffert S, Krämer R, Revuelta JL (1999) Physiological consequence of disruption of the *VMA1* gene in the riboflavin overproducer *Ashbya gossypii*. J Biol Chem 274:9442–9448. https://doi.org/10.1074/jbc.274.14.9442

Gerstenberger JP, Occhipinti P, Gladfelter AS (2012) Heterogeneity in mitochondrial morphology and membrane potential is independent of the nuclear division cycle in multinucleate fungal cells. Eukaryot Cell 11:353–367. https://doi.org/10.1128/ec.05257-11

Gibeaux R, Hoepfner D, Schlatter I, Antony C, Philippsen P (2013) Organization of organelles within hyphae of *Ashbya gossypii* revealed by electron tomography. Eukaryot Cell 12:1423–1432. https://doi.org/10.1128/ec.00106-13

Gladfelter AS, Hungerbuehler AK, Philippsen P (2006) Asynchronous nuclear division cycles in multinucleated cells. J Cell Biol 3:347–362. https://doi.org/10.1083/jcb.200507003

Gomes D, Aguiar TQ, Dias O, Ferreira EC, Domingues L, Rocha I (2014) Genome-wide metabolic re-annotation of *Ashbya gossypii*: New insights into its metabolism through a comparative analysis with *Saccharomyces cerevisiae* and *Kluyveromyces lactis*. BMC Genomics 15:810. https://doi.org/10.1186/1471-2164-15-810

Grava S, Philippsen P (2010) Dynamics of multiple nuclei in *Ashbya gossypii* hyphae depend on the control of cytoplasmic microtubules length by Bik1, Kip2, Kip3, and not on a capture/shrinkage mechanism. Mol Biol Cell 21:3680–3692. https://doi.org/10.1091/mbc.e10-06-0527

Grünler A, Walther A, Lämmel J, Wendland J (2010) Analysis of flocculins in *Ashbya gossypii* reveals *FIG2* regulation by *TEC1*. Fungal Genet Biol 47:619–628. https://doi.org/10.1016/j.fgb.2010.04.001

Helfer H, Gladfelter AS (2006) AgSwe1p regulates mitosis in response to morphogenesis and nutrients in multinucleated *Ashbya gossypii* cells. Mol Biol Cell 17:4494–4512. https://doi.org/10.1091/mbc.e06-03-0215

Hungerbuehler AK, Philippsen P, Gladfelter AS (2007) Limited functional redundancy and oscillation of cyclins in multinucleated *Ashbya gossypii* fungal cells. Eukaryot Cell 6:473–486. https://doi.org/10.1128/ec.00273-06

Jiménez A, Santos MA, Pompejus M, Revuelta JL (2005) Metabolic engineering of the purine pathway for riboflavin production in *Ashbya gossypii*. Appl Environ Microbiol 71:5743–5751. https://doi.org/10.1128/aem.71.10.5743-5751.2005

Jiménez A, Santos MA, Revuelta JL (2008) Phosphoribosyl pyrophosphate synthetase activity affects growth and riboflavin production in *Ashbya gossypii*. BMC Biotechnol 8:67–78. https://doi.org/10.1186/1472-6750-8-67

Jiménez A, Muñoz-Fernández G, Ledesma-Amaro R, Buey RM, Revuelta JL (2019) One-vector CRISPR/Cas9 genome engineering of the industrial fungus *Ashbya gossypii*. Microb Biotechnol 12:1293–1301. https://doi.org/10.1111/1751-7915.13425

Jiménez A, Hoff B, Revuelta JL (2020) Multiplex genome editing in *Ashbya gossypii* using CRISPR-Cpf1. N Biotechnol 57:29–33. https://doi.org/10.1016/j.nbt.2020.02.002

Käsler B, Sahm H, Stahmann KP, Schmidt G, Boeddecker B, Seulberger H (1997) Riboflavin production process by means of microorganisms with modified isocitrate lyase activity. WO Patent 1997003208 A1, 30 Jan 1997. https://patents.google.com/patent/WO1997003208A1/en

Kato T, Park EY (2006) Expression of alanine:glyoxylate aminotransferase gene from *Saccharomyces cerevisiae* in *Ashbya gossypii*. Appl Microbiol Biotechnol 71:46–52. https://doi.org/10.1007/s00253-005-0124-5

Kato T, Azegami J, Kano M, El Enshasy HA, Park EY (2021) Effects of sirtuins on the riboflavin production in *Ashbya gossypii*. Appl Microbiol Biotechnol 105:7813–7823. https://doi.org/10.1007/s00253-021-11595-2

Kato T, Kano M, Yokomori A, Azegami J, El Enshasy HA, Park EY (2023) Involvement of a flavoprotein, acetohydroxyacid synthase, in growth and riboflavin production in riboflavin-overproducing *Ashbya gossypii* mutant. Microb Cell Fact 22:105. https://doi.org/10.1186/s12934-023-02114-1

Kato T, Azegami J, Kano M, El Enshasy HA, Park EY (2024) Induction of oxidative stress in sirtuin gene-disrupted *Ashbya gossypii* mutants overproducing riboflavin. Mol Biotechnol 66:1144–1153. https://doi.org/10.1007/s12033-023-01012-6

Kaufmann A (2009) A plasmid collection for PCR-based gene targeting in the filamentous ascomycete *Ashbya gossypii*. Fungal Genet Biol 46:595–603. https://doi.org/10.1016/j.fgb.2009.05.002

Kavitha S, Chandra TS (2009) Effect of vitamin E and menadione supplementation on riboflavin production and stress parameters in *Ashbya gossypii*. Proc Bio 44:934–938. https://doi.org/10.1016/j.procbio.2009.04.001

Kavitha S, Chandra TS (2014) Oxidative stress protection and glutathione metabolism in response to hydrogen peroxide and menadione in riboflavinogenic fungus *Ashbya gossypii*. Appl Biochem Biotechnol 174:2307–2325. https://doi.org/10.1007/s12010-014-1188-4

Knechtle P, Dietrich F, Philippsen P (2003) Maximal polar growth potential depends on the polarisome component AgSpa2 in the filamentous fungus *Ashbya gossypii*. Mol Biol Cell 14:4140–4154. https://doi.org/10.1091/mbc.e03-03-0167

Lago BD, Kaplan L (1981) Vitamin fermentations: B2 and B12. In: Moo-Young M, Vezina C, Singh K (eds) Advances in biotechnology: fermentation products, vol 3. Pergamon Press, Oxford, pp 241–246. https://doi.org/10.1016/B978-0-08-025385-5.50045-6

Lang C, Grava S, van den Hoorn T, Trimble R, Philippsen P, Jaspersen SL (2010) Mobility, microtubule nucleation and structure of microtubule-organizing centers in multinucleated hyphae of *Ashbya gossypii*. Mol Biol Cell 21:18–28. https://doi.org/10.1091/mbc.e09-01-0063

Ledesma-Amaro R, Santos MA, Jiménez A, Revuelta JL (2014a) Strain design of *Ashbya gossypii* for single-cell oil production. Appl Environ Microbiol 80:1237–1244. https://doi.org/10.1128/aem.03560-13

Ledesma-Amaro R, Santos MA, Jiménez A, Revuelta JL (2014b) Tuning single-cell oil production in *Ashbya gossypii* by engineering the elongation and desaturation systems. Biotechnol Bioeng 111:1782–1791. https://doi.org/10.1002/bit.25245

Ledesma-Amaro R, Buey RM, Revuelta JL (2015a) Increased production of inosine and guanosine by means of metabolic engineering of the purine pathway in *Ashbya gossypii*. Microb Cell Fact 14:58. https://doi.org/10.1186/s12934-015-0234-4

Ledesma-Amaro R, Serrano-Amatriain C, Jiménez A, Revuelta JL (2015b) Metabolic engineering of riboflavin production in *Ashbya gossypii* through pathway optimization. Microb Cell Fact 14:163. https://doi.org/10.1186/s12934-015-0354-x

Ledesma-Amaro R, Buey RM, Revuelta JL (2016) The filamentous fungus *Ashbya gossypii* as a competitive industrial inosine producer. Biotechnol Bioeng 113:2060–2063. https://doi.org/10.1002/bit.25965

Ledesma-Amaro R, Jiménez A, Revuelta JL (2018) Pathway grafting for polyunsaturated fatty acids production in *Ashbya gossypii* through Golden Gate rapid assembly. ACS Synth Biol 7:2340–2347. https://doi.org/10.1021/acssynbio.8b00287

Lee C, Occhipinti P, Gladfelter AS (2015) PolyQ-dependent RNA-protein assemblies control symmetry breaking. J Cell Biol Mar 208:533–544. https://doi.org/10.1083/jcb.201407105

Lengeler KB, Wasserstrom L, Walther A, Wendland J (2013) Analysis of the cell wall integrity pathway of *Ashbya gossypii*. Microbiol Res 168:607–614. https://doi.org/10.1016/j.micres.2013.06.008

Lizama G, Moguel-Salazar F, Peraza-Luna F, Ortiz-Vázquez E (2007) Riboflavin production from mutants of *Ashbya gossypii* utilising orange rind as a substrate. Ann Microbiol 57:157–161. https://doi.org/10.1007/BF03175201

Lozano-Martínez P, Buey RM, Ledesma-Amaro R, Jiménez A, Revuelta JL (2017) Engineering *Ashbya gossypii* strains for *de novo* lipid production using industrial by-products. Microb Biotechnol 10:425–433. https://doi.org/10.1111/1751-7915.12487

Maeting I, Schmidt G, Sahm H, Revuelta JL, Stierhofd YD, Stahmanna KP (1999) Isocitrate lyase of *Ashbya gossypii*—transcriptional regulation and peroxisomal localization. FEBS Lett 444:15–21. https://doi.org/10.1016/s0014-5793(99)00017-4

Magalhães F, Aguiar TQ, Oliveira C, Domingues L (2014) High-level expression of *Aspergillus niger β*-galactosidase in *Ashbya gossypii*. Biotechnol Prog 30:261–268. https://doi.org/10.1002/btpr.1844

Mateos L, Jiménez A, Revuelta J, Santos M (2006) Purine biosynthesis, riboflavin production, and trophic phase span are controlled by a Myb-related transcription factor in the fungus *Ashbya gossypii*. Appl Environ Microbiol 72:5052–5060. https://doi.org/10.1128/aem.00424-06

Monschau N, Sahm H, Stahmann K (1998) Threonine aldolase overexpression plus threonine supplementation enhanced riboflavin production in *Ashbya gossypii*. Appl Environ Microbiol 64:4283–4290. https://doi.org/10.1128/aem.64.11.4283-4290.1998

Montero-Bullón JF, Martín-González J, Ledesma-Amaro R, Jiménez A, Buey RM (2025a) Pioneering microbial synthesis of gangliosides in the filamentous fungus *Ashbya gossypii*. Biotechnol Biofuels Bioprod 18:95. https://doi.org/10.1186/s13068-025-02697-4

Montero-Bullón JF, Martín-González J, Muñoz-Fernández G, Buey RM, Jiménez A (2025b) Lipid metabolism optimisation in *Ashbya gossypii* to produce microbial oils from xylose-rich feedstocks. Microb Biotechnol 18:e70209. https://doi.org/10.1111/1751-7915.70209

Muñoz-Fernández G, Montero-Bullón JF, Revuelta JL, Jiménez A (2021) New promoters for metabolic engineering of *Ashbya gossypii*. J Fungi (Basel) 7:906. https://doi.org/10.3390/jof7110906

Muñoz-Fernández G, Martínez-Buey R, Revuelta JL, Jiménez A (2022) Metabolic engineering of *Ashbya gossypii* for limonene production from xylose. Biotechnol Biofuels Bioprod 15:79. https://doi.org/10.1186/s13068-022-02176-0

Muñoz-Fernández G, Montero-Bullón JF, Martínez JL, Buey RM, Jiménez A (2024) *Ashbya gossypii* as a versatile platform to produce sabinene from agro-industrial wastes. Fungal Biol Biotechnol 11:16. https://doi.org/10.1186/s40694-024-00186-1

Muñoz-Fernández G, Jiménez A, Revuelta JL (2020) Genomic edition of *Ashbya gossypii* using one-vector CRISPR/Cas9. Bio Protoc 10:e3660. https://doi.org/10.21769/bioprotoc.3660

Nieland S, Stahmann KP (2013) A developmental stage of hyphal cells shows riboflavin overproduction instead of sporulation in *Ashbya gossypii*. Appl Microbiol Biotechnol 97:10143–10153. https://doi.org/10.1007/s00253-013-5266-2

Nordmann D, Lickfeld M, Warnsmann V, Wiechert J, Jendretzki A, Schmitz HP (2014) The small GTP-binding proteins AgRho2 and AgRho5 regulate tip-branching, maintenance of the growth axis and actin-ring-integrity in the filamentous fungus *Ashbya gossypii*. PLoS ONE 9:e106236. https://doi.org/10.1371/journal.pone.0106236

Oliveira C, Aguiar TQ, Domingues L (2017) Principles of genetic rngineering. In: Pandey A, Teixeira JAC (eds) Current developments in biotechnology and bioengineering: foundations of biotechnology and bioengineering. Elsevier, Amsterdam, pp 81–127. https://doi.org/10.1016/B978-0-444-63668-3.00004-4

Park EY, Zhang JH, Tajima S, Dwiarti L (2007) Isolation of *Ashbya gossypii* mutant for an improved riboflavin production targeting for biorefinery technology. J Appl Microbiol 103:468–476. https://doi.org/10.1111/j.1365-2672.2006.03264.x

Park EY, Ito Y, Nariyama M, Sugimoto T, Lies D, Kato T (2011) The improvement of riboflavin production in *Ashbya gossypii* via disparity mutagenesis and DNA microarray analysis. Appl Microbiol Biotechnol 91:1315–1326. https://doi.org/10.1007/s00253-011-3325-0

Patel MV, Chandra TS (2020) Metabolic engineering of *Ashbya gossypii* for enhanced FAD production through promoter replacement of *FMN1* gene. Enzyme Microb Technol 133:109455. https://doi.org/10.1016/j.enzmictec.2019.109455

Perlman D (1979) Microbial process for riboflavin production. In: Pepplev HJ, Pevlman D (eds) Microbial technology, vol 1. Academic, New York, pp 521–527. https://doi.org/10.1016/B978-0-12-551501-6.50021-8

Pfeifer VF, Tanner FW, Vojnovich C, Traufler DH (1950) Riboflavin by fermentation with *Ashbya gossypii*. Ind Eng Chem 42:1776–1781. https://doi.org/10.1021/ie50489a027

Pridham TG, Raper KB (1952) Studies on variation and mutation in *Ashbya gossypii*. Mycologia 44:452–469. https://doi.org/10.1080/00275514.1952.12024209

Ravasio D, Wendland J, Walther A (2014) Major contribution of the Ehrlich pathway for 2-phenylethanol/rose flavor production in *Ashbya gossypii*. FEMS Yeast Res 14:833–844. https://doi.org/10.1111/1567-1364.12172

Revuelta JL, Ledesma-Amaro R, Lozano-Martinez P, Díaz-Fernández D, Buey RM, Jiménez A (2017) Bioproduction of riboflavin: a bright yellow history. J Ind Microbiol Biotechnol 44:659–665. https://doi.org/10.1007/s10295-016-1842-7

Revuelta JL, Serrano-Amatriain C, Ledesma-Amaro R, Jiménez A (2018) Formation of folates by microorganisms: towards the biotechnological production of this vitamin. Appl Microbiol Biotechnol 102:8613–8620. https://doi.org/10.1007/s00253-018-9266-0

Revuelta JL, Santos MA, Pompejus M, Seulberger H (1999) Promoter from *Ashbya gossypii*. WO Patent 1999033993 A1, 8 Jul 1999. https://patents.google.com/patent/WO1999033993A1/en

Ribeiro O, Wiebe M, Ilmén M, Domingues L, Penttilä M (2010) Expression of *Trichoderma reesei* cellulases CBHI and EGI in *Ashbya gossypii*. Appl Microbiol Biotechnol 87:1437–1446. https://doi.org/10.1007/s00253-010-2610-7

Ribeiro O, Domingues L, Penttilä M, Wiebe MG (2012) Nutritional requirements and strain heterogeneity in *Ashbya gossypii*. J Basic Microbiol 52:582–589. https://doi.org/10.1002/jobm.201100383

Ribeiro O, Magalhães F, Aguiar TQ, Wiebe MG, Penttilä M, Domingues L (2013) Random and direct mutagenesis to enhance protein secretion in *Ashbya gossypii*. Bioengineered 4:1–10. https://doi.org/10.4161/bioe.24653

Santos MA, Mateos L, Stahmann KP, Revuelta JL (2005) Insertional mutagenesis in the vitamin B2 producer fungus *Ashbya gossypii*. In: Barredo JL (ed) Microbial processes and products: methods in biotechnology, vol 18. Humana Press, New Jersey, pp 283–300. https://doi.org/10.1385/1-59259-847-1:283

Schlösser T, Schmidt G, Stahmann KP (2001) Transcriptional regulation of 3,4-dihydroxy-2-butanone 4-phosphate synthase. Microbiology 147:3377–3386. https://doi.org/10.1099/00221287-147-12-3377

Schlösser T, Wiesenburg A, Gätgens C, Funke A, Viets U, Vijayalakshmi S, Nieland S, Stahmann KP (2007) Growth stress triggers riboflavin overproduction in *Ashbya gossypii*. Appl Microbiol Biotechnol 76:569–578. https://doi.org/10.1007/s00253-007-1075-9

Schlüpen C, Santos MA, Weber U, de Graaf A, Revuelta JL, Stahmann KP (2003) Disruption of the *SHM2* gene, encoding one of two serine hydroxymethyltransferase isozymes, reduces the flux from glycine to serine in *Ashbya gossypii*. Biochem J 369:263–273. https://doi.org/10.1042/bj20021224

Schlüpen C (2003) Untersuchungen zur bedeutung verschiedener enzyme des glycin-stoffwechsels für die riboflavin-bildung in *Ashbya gossypii*. PhD Thesis. Heinrich Heine University of Düsseldorf, Germany, pp 118. https://docserv.uni-duesseldorf.de/servlets/DocumentServlet?id=2656

Schmidt G, Stahmann KP, Sahm H (1996) Inhibition of purified isocitrate lyase identified itaconate and oxalate as potential antimetabolites for the riboflavin overproducer *Ashbya gossypii*. Microbiology 142:411–417. https://doi.org/10.1099/13500872-142-2-411

Semenova E, Presniakova V, Kozlovskaya V, Markelova N, Gusev A, Linert W, Kurakov A, Shpichka A (2022) The *in vitro* cytotoxicity of *Eremothecium* oil and its components—aromatic and acyclic monoterpene alcohols. Int J Mol Sci 23:3364. https://doi.org/10.3390/ijms23063364

Serrano-Amatriain C, Ledesma-Amaro R, López-Nicolás R, Ros G, Jiménez A, Revuelta JL (2016) Folic acid production by engineered *Ashbya gossypii*. Metab Eng 38:473–482. https://doi.org/10.1016/j.ymben.2016.10.011

Silva R, Aguiar TQ, Domingues L (2015) Blockage of the pyrimidine biosynthetic pathway affects riboflavin production in *Ashbya gossypii*. J Biotechnol 193:37–40. https://doi.org/10.1016/j.jbiotec.2014.11.009

Silva R, Aguiar TQ, Coelho E, Jiménez A, Revuelta JL, Domingues L (2019a) Metabolic engineering of *Ashbya gossypii* for deciphering the *de novo* biosynthesis of γ-lactones. Microb Cell Fact 18:62. https://doi.org/10.1186/s12934-019-1113-1

Silva R, Aguiar TQ, Oliveira C, Domingues L (2019b) Physiological characterization of a pyrimidine auxotroph exposes link between uracil phosphoribosyltransferase regulation and riboflavin production in *Ashbya gossypii*. N Biotechnol 50:1–8. https://doi.org/10.1016/j.nbt.2018.12.004

Silva R, Coelho E, Aguiar TQ, Domingues L (2021) Microbial biosynthesis of lactones: Gaps and opportunities towards sustainable production. Appl Sci 11:8500. https://doi.org/10.3390/APP11188500

Silva R, Aguiar TQ, Domingues L (2022) Orotic acid production from crude glycerol by engineered *Ashbya gossypii*. Bioresour Technol Rep 17:100992. https://doi.org/10.1016/j.biteb.2022.100992

Silva R, Aguiar TQ, Oliveira R, Domingues L (2019c) Light exposure during growth increases riboflavin production, reactive oxygen species accumulation and DNA damage in *Ashbya gossypii* riboflavin-overproducing strains. FEMS Yeast Res 19:foy114. https://doi.org/10.1093/femsyr/foy114

Silva R (2014) Exploitation of *Ashbya gossypii* for the production of high-value products from glycerol feedstocks. MSc Thesis, University of Minho. https://hdl.handle.net/1822/41689

Stahmann KP, Kupp C, Feldmann SD, Sahm H (1994) Formation and degradation of lipid bodies found in the riboflavin-producing fungus *Ashbya gossypii*. Appl Microbiol Biotechnol 42:121–127. https://doi.org/10.1007/BF00170234

Steiner S, Philippsen P (1994) Sequence and promoter analysis of the highly expressed *TEF* gene of the filamentous fungus *Ashbya gossypii*. Mol Gen Genet 242:263–271. https://doi.org/10.1007/BF00280415

Steiner S, Wendland J, Wright MC, Philippsen P (1995) Homologous recombination as the main mechanism for DNA integration and cause of rearrangements in the filamentous ascomycete *Ashbya gossypii*. Genetics 140:973–987. https://doi.org/10.1093/genetics/140.3.973

Sugimoto T, Kanamasa S, Kato T, Park EY (2009) Importance of malate synthase in the glyoxylate cycle of *Ashbya gossypii* for the efficient production of riboflavin. Appl Microbiol Biotechnol 83:529–539. https://doi.org/10.1007/s00253-009-1972-1

Sugimoto T, Morimoto A, Nariyama M, Kato T, Park EY (2010) Isolation of an oxalate-resistant *Ashbya gossypii* strain and its improved riboflavin production. J Ind Microbiol Biotechnol 37:57–64. https://doi.org/10.1007/s10295-009-0647-3

Tajima S, Itoh Y, Sugimoto T, Kato T, Park EY (2009) Increased riboflavin production from activated bleaching earth by a mutant strain of *Ashbya gossypii*. J Biosci Bioeng 108:325–329. https://doi.org/10.1016/j.jbiosc.2009.04.021

Wabner D, Overhageböck T, Nordmann D, Kronenberg J, Kramer F (2019) Schmitz HP (2019) Analysis of the protein composition of the spindle pole body during sporulation in *Ashbya gossypii*. PLoS ONE 14:e0223374. https://doi.org/10.1371/journal.pone.0223374

Wach A, Brachat A, Pöhlmann R, Philippsen P (1994) New heterologous modules for classical or PCR-based gene disruptions in *Saccharomyces cerevisiae*. Yeast 10:1793–1808. https://doi.org/10.1002/yea.320101310

Walther A, Wendland J (2012) Yap1-dependent oxidative stress response provides a link to riboflavin production in *Ashbya gossypii*. Fungal Genet Biol 49:697–707. https://doi.org/10.1016/j.fgb.2012.06.006

Wasserstrom L, Lengeler K, Walther A, Wendland J (2015) Developmental growth control exerted via the protein A kinase Tpk2 in *Ashbya gossypii*. Eukaryot Cell 14:593–601. https://doi.org/10.1128/ec.00045-15

Wasserstrom L, Dünkler A, Walther A, Wendland J (2017) The APSES protein Sok2 is a positive regulator of sporulation in *Ashbya gossypii*. Mol Microbiol 106:949–960. https://doi.org/10.1111/mmi.13859

Wei S, Hurley J, Jiang Z, Wang S, Wang Y (2012) Isolation and characterization of an *Ashbya gossypii* mutant for improved riboflavin production. Braz J Microbiol 43:441–448. https://doi.org/10.1590/s1517-83822012000200003

Wendland J (2003) Analysis of the landmark protein Bud3 of *Ashbya gossypii* reveals a novel role in septum construction. EMBO Rep 4:200–204. https://doi.org/10.1038/sj.embor.embor727

Wendland J, Ayad-Durieux Y, Knechtle P, Rebischung C, Philippsen P (2000) PCR-based gene targeting in the filamentous fungus *Ashbya gossypii*. Gene 242:381–391. https://doi.org/10.1016/s0378-1119(99)00509-0

Wendland J, Dünkler A, Walther A (2011) Characterization of α-factor pheromone and pheromonereceptor genes of *Ashbya gossypii*. FEMS Yeast Res 11:418–429. https://doi.org/10.1111/j.1567-1364.2011.00732.x

Wright MC, Philippsen P (1991) Replicative transformation of the filamentous fungus *Ashbya gossypii* with plasmids containing *Saccharomyces cerevisiae* ARS elements. Gene 109:99–105. https://doi.org/10.1016/0378-1119(91)90593-z

Chapter 4
Bioprocess and Biosustainable Strategies

Abstract This chapter provides a comprehensive review on the development of *A. gossypii* bioprocesses for the biosynthesis of metabolites of biotechnological interest. The first subsection focuses on bioprocess development, mostly focusing on riboflavin production, the main staple in *Ashbya* biotechnology. It covers the evolution of *A. gossypii* bioprocesses in what regards to the establishment and impact of nutritional requirements and media formulation, definition of process parameters (pH, temperature, inoculum), process configuration (e.g. batch and fed-batch), scaling up and downstream processing. The second subsection focuses on the development of biosustainable strategies mainly translated by the usage of low-cost, renewable and sustainable feedstocks, such as agro-industrial by-products and wastes. In this section, alternative nitrogen and carbon sources are explored, with particular emphasis on emerging substrates such as oil wastes, lignocellulosic residues and crude glycerol and their application to produce riboflavin as well as emergent value-added metabolites such as microbial oils, terpenes and orotic acid.

Keywords *Ashbya gossypii* · Bioprocess optimization · Media formulation · Nitrogen sources · Carbon sources · Low-cost substrates · Agro-industrial by-products · Sustainable bioproduction

4.1 Bioprocess Optimization

Efforts to understand and optimize *A. gossypii* growth and metabolism date as far back as 1930, with works like the one conducted by Farries and Bell establishing early indications on *A. gossypii*'s nutritional requirements. Early on, these identified growth performance in various carbon sources, namely hexoses (glucose, fructose, mannose, rhamnose, galactose, and mannitol), pentoses (arabinose and xylose), disaccharides (cane sugar, maltose, and lactose) and starch. Hexoses were shown to be the most favorable sugars, whereas pentoses supported only negligible growth. The fungus was also able to metabolize dimers such as sucrose and maltose, but not lactose. In addition, various nitrogen sources were evaluated, including complex protein sources (peptone, egg white, gluten, casein, animal protein, and gelatin), synthetic sources (single amino acids and amino acid mixtures), and inorganic nitrogen compounds (potassium nitrate, ammonium sulfate and ammonium chloride). Complex nitrogen sources were the most favorable substrates, particularly when hydrolyzed (Farries and Bell 1930). Apart from macronutrients other works identified additional nutritional requirements, particularly regarding vitamins. As reviewed by Pridham and Rapper, in more defined nitrogen sources such as amino acid mixtures, *A. gossypii* requires several accessory factors for proper growth, such as myo-inositol, biotin and thiamin, with the latter being highlighted as particularly important (Pridham and Raper 1950).

Early on, Farries and Bell mentioned in their work a yellow coloration in some of the conditions, later known to be associated with riboflavin production. With this knowledge, and already envisioning a commercial product, Tanner et al. developed efforts to maximize riboflavin production through optimization of fermentation conditions. This work identified an optimal production temperature range of 26–28 °C, a pH between 6–7 and 3–4% of glucose concentration, setting the base for the optimal bioprocess conditions. Inoculum and medium preparation procedures were also evaluated, with younger inocula (24 h) showing superior performance. In agreement with previous findings, Tanner et al. also pointed out complex nitrogen sources, such as corn steep liquor, animal protein and commercial peptones, as the most favorable for riboflavin production (Tanner et al. 1949).

Building on these results, Pfeifer et al. (1950) transferred the process to pilot scale and examined the impact of media composition, inoculum preparation and aeration on riboflavin production, achieving titers between 500 and 848 µg/mL. Translation of the process into pilot scale was performed in a 4000-gallon fermenter, with a 1250-gallon working volume, including downstream processing steps as evaporation and drying, which enabled a 96% recovery efficiency of the produced riboflavin (Pfeifer et al. 1950).

Riboflavin yields were pushed further by identifying oil as a beneficial carbon and energy source for riboflavin overproduction in *A. gossypii*, reaching titers of about 4 mg/mL when using corn steep liquor, corn oil, glycine and enzymatically degraded collagen (Malzahn et al. 1959). Optimal fermentation conditions, as reviewed by Perlman, were achieved with 28 °C, approximately 0.3 vvm of aeration and agitation with three impellers at 1 hp/1000 L (Perlman 1979).

The importance of oil as a carbon source and its positive impact on riboflavin overproduction was further studied by Özbas and Kutsal (1986), who demonstrated a two- to threefold increase in specific productivity through the partial replacement of glucose with oil (Özbas and Kutsal 1986). Additionally, Özbas and Kutsal corroborated the importance of vitamins like biotin, thiamin and myo-inositol previously mentioned, in the case of fermentations using glucose or whey as substrate, but reported a dissimilar effect in fermentations with oil, having reported a negative impact on riboflavin production upon addition of these vitamins (Özbaş and Kutsal 1991).

Fermentations combining glucose and oil were further developed by Soo-Wan et al. (1993), who optimized the process and implemented a fed-batch configuration,

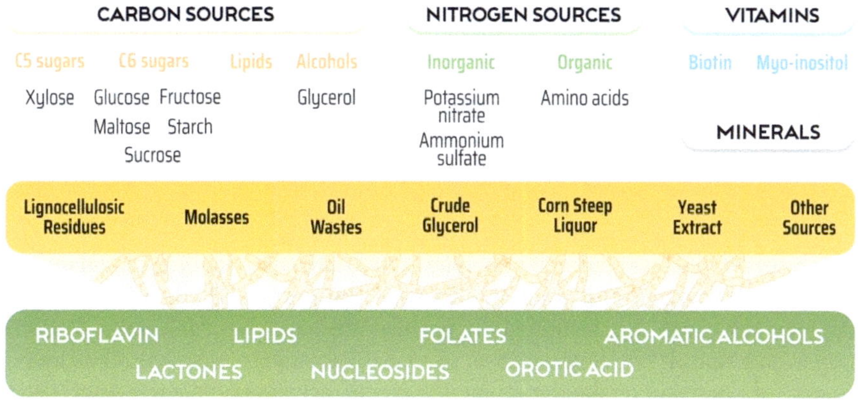

Fig. 4.1 Compilation of the main substrates, either pure or complex, and the main products deriving from *Ashbya gossypii* based bioprocesses

achieving riboflavin titers of 5.4 g/L in batch configuration, which increased to 6.8 g/L under the fed-batch approach (Soo-Wan et al. 1993).

With the strong knowledge base established and the emergence of molecular tools enabling genetic modification of *A. gossypii*, most optimization efforts shifted toward metabolic engineering approaches, as discussed in the previous chapter, while less emphasis was placed on bioprocess engineering. The discovery of novel metabolites of biotechnological interest allowed much of the knowledge previously acquired in riboflavin production to serve as a foundation for developing bioprocesses targeting these new compounds (Fig. 4.1).

4.2 Low-Cost Substrates

One important aspect to consider when establishing a biotechnological product is the overall techno-economic viability and sustainability of the production process. When considering fermentation-based products, an important factor to take into account is the type and cost of substrates used, which accounts for a significant portion of the fermentation associated costs (Pereira et al. 2025). As a result, several works have put a particular focus on the pursuit and evaluation of lower cost substrates for bioproduction, envisaging not only cost reduction but also the fostering of a bio-based circular economy (Francisco et al. 2023; Pereira et al. 2025).

As mentioned previously, early works have characterized a wide range of substrates for the supply of nitrogen, including corn steep liquor, meat scrap residues, gelatin and other renewable sources (Farries and Bell 1930; Pridham and Raper 1950), which were steadily used in several of the previously mentioned optimization works. Throughout these, corn steep liquor became one of the most used substrates, representing a preferred source for *A. gossypii* fermentations and the production of diverse metabolites such as riboflavin (Park et al. 2007), orotic acid (Silva et al. 2022) and single cell oils (Francisco et al. 2023). This residue derives from the wet milling process in corn starch refining and supplies several of the required nutrients, such as nitrogen, vitamins, and salts (Jiao et al. 2022). With the establishment of preferred nitrogen sources, research focused on alternative low-cost carbon sources, which are often considered to be the main contributor to operational costs, especially in high yield fermentation processes (Hepner 1996).

One early work focusing on waste utilization was developed by Ming et al. (2003), who resorted to waste activated bleaching earth (ABE) deriving from an oil refinery plant, which contains 40% of oil on a dry basis (Ming et al. 2003). Following up on the previous works that identified the positive impact of oil (Malzahn et al. 1959; Özbas and Kutsal 1986), Ming et al. used ABE to supply rapeseed oil as carbon source for riboflavin production using *A. gossypii*, achieving titers of 1.1 g/L (Ming et al. 2003). This work was further developed by Park et al. (2004), also using waste ABE but this time containing palm oil residues, achieving a higher riboflavin titer of 2.1 g/L (Park et al. 2004). Later on, Tajima et al. (2009), combined this strategy with metabolically engineered strains and pushed riboflavin titers even further to 8.7 g/L (Tajima et al. 2009). In recent works, and using directly oil wastes as substrate, in this case cooking oil, Martín-González et al. demonstrated in 2025 the feasibility of using these residues to produce a wider variety of metabolites using genetically engineered *A. gossypii* strains to obtain riboflavin (312.5 mg/L), folates (7.6 mg/L), biolipids (80% of lipid accumulation) and monoterpenes like limonene and sabinene (140 mg/L with oil, increased to 300 mg/L with xylose supplementation) (Martín-González et al. 2025).

Apart from residues containing lipids, other byproducts were also explored, focusing on the supply of sugars as carbon source. A notable example is molasses, an abundant byproduct of sugarcane or beet processing in sugar refining, which contains high concentrations of sucrose, along with glucose and fructose (Huang et al. 2013).

Lozano-Martínez et al. (2017) used sugarcane molasses targeting *de novo* production of lipids by *A. gossypii*, achieving accumulation of lipids in the biomass of around 25% of the dry cell weight (Lozano-Martínez et al. 2017). This work was followed up by Díaz-Fernández et al. (2019), that proposed co-utilization approaches by combining different wastes, such as lignocellulosic hydrolysates, and molasses or crude glycerol (Díaz-Fernández et al. 2019).

Lignocellulosic residues are common by-products of agro-industrial activity and an abundant carbon source, representing an opportunity for biotechnological applications envisaging a sustainable circular economy in the context of biorefineries (Romaní et al. 2019). These are rich in cellulose and hemicellulose that can be hydrolyzed into C6 and C5 sugars, predominantly glucose and xylose that can be metabolized into value added products (Francisco et al. 2023).

In their work in 2019, Díaz-Fernández et al. applied genetically modified *A. gossypii* strains in fermentations using corn cob hydrolysates, combined either with sugarcane molasses or with crude glycerol for the production of microbial lipids, yielding a maximum lipid titer of 3.9 g/L and an accumulation of about 40% of dry cell weight in lipids (Díaz-Fernández et al. 2019).

This type of strategy was also utilized for producing monoterpenes like limonene and sabinene by Muñoz-Fernández et al. (2024), complementing the previously mentioned work of Martín-González published in 2025. In their approach, Muñoz-Fernández et al. combined corn cob hydrolysates with either beet or sugarcane molasses and applied genetically modified *A. gossypii* strains to produce monoterpenes, achieving limonene titers of 383 mg/L and sabinene titers of 685 mg/L (Muñoz-Fernández et al. 2024).

A more focused approach was undertaken by Francisco et al., who focused on the sole use of lignocellulosic residues for carbon supply. This approach resorted to eucalyptus bark hydrolysates to produce microbial lipids using engineered *A. gossypii* strains, achieving a maximum lipid titer of 1.42 g/L using non-detoxified media in a bioreactor configuration (Francisco et al. 2023).

Another example of the use of an industrial waste for carbon supply is crude glycerol, a by-product of biodiesel production. This residue was already mentioned previously in the context of co-utilization approaches, having been used in combination with hydrolyzed corn cob to produce microbial lipids in the work of Días-Fernández et al. (2019). This residue was further explored by Silva et al. (2022) work, that used it as the sole carbon source to produce orotic acid with genetically modified *A. gossypii* strains, achieving titers of about 4 g/L, slightly higher when compared to those obtained with synthetic/pure glycerol (Silva et al. 2022).

References

Díaz-Fernández D, Aguiar TQ, Martín VI, Romaní A, Silva R, Domingues L, Revuelta JL (2019) Microbial lipids from industrial wastes using xylose-utilizing *Ashbya gossypii* strains. Bioresour Technol 293:122054. https://doi.org/10.1016/J.BIORTECH.2019.122054

Farries EHM, Bell F (1930) On the metabolism of *Nematospora gossypii* and related fungi, with special reference to the source of nitrogen. Ann Bot os-44:423–455. https://doi.org/10.1093/OXFORDJOURNALS.AOB.A090228/2/OS-44-2-423.PDF.GIF

Francisco M, Aguiar TQ, Abreu G, Marques S, Gírio F, Domingues L (2023) Single-cell oil production by engineered *Ashbya gossypii* from non-detoxified lignocellulosic biomass hydrolysate. Fermentation 9:791. https://doi.org/10.3390/FERMENTATION9090791

Hepner L (1996) Cost analysis of fermentation processes. Chimia (Aarau) 50:442. https://doi.org/10.2533/CHIMIA.1996.442

Huang C, Chen X fang, Xiong L, Chen XD, Ma LL, Chen Y (2013) Single cell oil production from low-cost substrates: the possibility and potential of its industrialization. Biotechnol Adv 31:129–139. https://doi.org/10.1016/J.BIOTECHADV.2012.08.010

Jiao Y, Chen HD, Han H, Chang Y (2022) Development and utilization of corn processing by-products: a review. Foods 11:3709. https://doi.org/10.3390/FOODS11223709

Lozano-Martínez P, Buey RM, Ledesma-Amaro R, Jiménez A, Revuelta JL (2017) Engineering *Ashbya gossypii* strains for *de novo* lipid production using industrial by-products. Microb Biotechnol 10:425–433. https://doi.org/10.1111/1751-7915.12487

Malzahn RC, Phillips RF, Hanson AM (1959) Riboflavin process. US Patent 2,876,169, 3 May 1959. https://patents.google.com/patent/US2876169A/en

Martín-González J, Montero-Bullón JF, Muñoz-Fernández G, Buey RM, Jiménez A (2025) Valorization of waste cooking oil for bioproduction of industrially-relevant metabolites in *Ashbya gossypii*. N Biotechnol 88:32–38. https://doi.org/10.1016/J.NBT.2025.04.002

Ming H, Lara Pizarro AV, Park EY (2003) Application of waste activated bleaching earth containing rapeseed oil on riboflavin production in the culture of *Ashbya gossypii*. Biotechnol Prog 19:410–417. https://doi.org/10.1021/BP0257325

Muñoz-Fernández G, Montero-Bullón JF, Martínez JL, Buey RM, Jiménez A (2024) *Ashbya gossypii* as a versatile platform to produce sabinene from agro-industrial wastes. Fungal Biol Biotechnol 11:1–10. https://doi.org/10.1186/S40694-024-00186-1/TABLES/2

Özbas T, Kutsal T (1986) Comparative study of riboflavin production from two microorganisms: *Eremothecium ashbyii* and *Ashbya gossypii*. Enzyme Microb Technol 8:593–596. https://doi.org/10.1016/0141-0229(86)90116-X

Özbaş T, Kutsal T (1991) Effects of growth factors on riboflavin production by *Ashbya gossypii*. Enzyme Microb Technol 13:594–596. https://doi.org/10.1016/0141-0229(91)90096-S

Park EY, Kato A, Ming H (2004) Utilization of waste activated bleaching earth containing palm oil in riboflavin production by *Ashbya gossypii*. J Amer Oil Chem Soc 81:57–62. https://doi.org/10.1007/S11746-004-0857-Z

Park EY, Zhang JH, Tajima S, Dwiarti L (2007) Isolation of *Ashbya gossypii* mutant for an improved riboflavin production targeting for biorefinery technology. J Appl Microbiol 103:468–476. https://doi.org/10.1111/J.1365-2672.2006.03264.X

Pereira AA, Yaverino-Gutierrez MA, Monteiro MC, Souza BA, Bachheti RK, Chandel AK (2025) Precision fermentation in the realm of microbial protein production: State-of-the-art and future insights. Food Res Int 200:115527. https://doi.org/10.1016/J.FOODRES.2024.115527

Perlman D (1979) Microbial process for riboflavin production. In: Pepplev HJ, Pevlman D (eds) Microbial technology, vol 1. Academic, New York, pp 521–527. https://doi.org/10.1016/B978-0-12-551501-6.50021-8

Pfeifer VF, Tanner FW, Vojnovich C, Traufler DH (1950) Riboflavin by fermentation with *Ashbya gossypii*. Ind Eng Chem 42:1776–1781. https://doi.org/10.1021/IE50489A027

Pridham TG, Raper KB (1950) *Ashbya gossypii*—its significance in nature and in the laboratory. Mycologia 42:603–623. https://doi.org/10.1080/00275514.1950.12017863

Romaní A, Larramendi A, Yáñez R, Cancela A, Sánchez A, Teixeira JA, Domingues L (2019) Valorization of *Eucalyptus nitens* bark by organosolv pretreatment for the production of advanced biofuels. Ind Crops Prod 132:327–335. https://doi.org/10.1016/J.INDCROP.2019.02.040

Silva R, Aguiar TQ, Domingues L (2022) Orotic acid production from crude glycerol by engineered *Ashbya gossypii*. Bioresour Technol Rep 17.https://doi.org/10.1016/j.biteb.2022.100992

Soo-Wan N, Hyung-Wook J, Jae-Gu P, Kim IH, Mheen TI (1993) Media optimization and fed-batch fermentation for riboflavin overproduction by *Ashbya gossypii*. Kor J Appl Microbiol Biotechnol 21:615–621

Tajima S, Itoh Y, Sugimoto T, Kato T, Park EY (2009) Increased riboflavin production from activated bleaching earth by a mutant strain of *Ashbya gossypii*. J Biosci Bioeng 108:325–329. https://doi.org/10.1016/J.JBIOSC.2009.04.021

Tanner FW, Vojnovich C, Van Lanen JM (1949) Factors affecting riboflavin production by *Ashbya gossypii*. J Bacteriol 58:737–745. https://doi.org/10.1128/JB.58.6.737-745.1949

Chapter 5
Perspectives

Abstract *Ashbya gossypii* has emerged as a versatile and robust microbial platform with unique features that support its growing role in modern biotechnology. This chapter presents forward-looking perspectives on the organism, highlighting its natural metabolic richness, filamentous growth, and genetic proximity to *Saccharomyces cerevisiae*, which together provide a valuable framework for both fundamental research and industrial applications. Advances in molecular tools, including regulatory elements, CRISPR-based genome editing, and secretion optimization, are expected to enhance precision engineering and strain development. The chapter emphasizes the importance of selecting suitable host chassis early in the bioprocess, informed by the endogenous metabolic capacities of diverse *A. gossypii* isolates. Emerging strategies for co-production of multiple metabolites and co-valorization of diverse side-streams hold great promise for improving the techno-economic performance of industrial processes, while the integration of upstream and downstream operations supports scalable and sustainable bioproduction. Finally, regulatory considerations and safe implementation in food, feed, and pharmaceutical applications are discussed. Collectively, these perspectives position *A. gossypii* as a model organism and a cornerstone for circular, sustainable, and innovative bio-based production systems.

Keywords *Ashbya gossypii* · Microbial platform · Biotechnological potential · Bioprocess engineering · Co-production strategies · Sustainable bioproduction · Circular bioeconomy

A. gossypii stands as a remarkable case in modern biotechnology, distinguished by a unique set of features that this book has brought together:

- Recognized as a model organism, extensively studied for its characteristic filamentous growth.
- One of the earliest success stories where a biotechnological process outperformed and ultimately replaced a chemical-based production route.

T. Q. Aguiar et al., *Ashbya gossypii in Biotechnology*,
SpringerBriefs in Molecular Science, https://doi.org/10.1007/978-3-032-12435-7_5

- Closely related, at the phylogenetic level, to one of the most established industrial microbial hosts: the yeast *S. cerevisiae*.
- Supported by a strong knowledge base and a deep understanding of its genome and metabolism.
- Equipped with a wide range of molecular tools for genetic modification, with a proven track record of effectiveness in targeted metabolite production.
- Characterized by a rich endogenous metabolism, naturally producing a variety of valuable metabolites, further expanded through genetic and metabolic engineering.
- Notable for its metabolic flexibility, including the ability to utilize diverse substrates and withstand industrially relevant inhibitory conditions.

Together, these features reinforce *A. gossypii*'s position as a robust and versatile production platform for biotechnological applications, keeping this fungus in the spotlight as a source of new knowledge and innovative solutions, positioning it as a valuable contributor to the future bioeconomy and the transition toward sustainable, circular production systems.

Hyphal growth and morphology in *A. gossypii* remain only partially understood, leaving important questions open for fundamental research. Continued investigation in this area is expected to expand our knowledge of both the evolutionary trajectory and the physiological mechanisms underlying this fungus. Of particular relevance is its close phylogenetic relationship and genetic similarity to *S. cerevisiae*, which provides a unique comparative framework. While *S. cerevisiae* is a well-established unicellular model, *A. gossypii* offers valuable insights into the transition to and regulation of filamentous growth. Studying its growth patterns, morphogenesis, and multinucleated hyphae therefore has the potential not only to deepen our understanding of fungal biology but also to shed light on the broader principles governing eukaryotic cell differentiation, development, and adaptation.

Beyond fundamental science, unraveling the mechanisms of hyphal growth holds clear biotechnological relevance. Morphology directly impacts nutrient uptake, metabolite secretion, and biomass distribution in large-scale cultivation processes. A deeper mechanistic understanding could enable the rational engineering of strains with optimized growth forms, tailored for improved productivity, robustness, and scalability in industrial fermentation. Thus, advancing knowledge in this field is expected to translate into both scientific insights and practical innovations, reinforcing *A. gossypii*'s potential as a versatile microbial cell factory.

The development of molecular tools in biotechnology is steadily advancing toward greater precision, versatility, and scalability, and *A. gossypii* is expected to follow this general trend. As the field embraces increasingly refined approaches to genetic and metabolic engineering, the molecular toolbox for this filamentous fungus will expand rapidly, creating new opportunities for fine-tuned strain design and reinforcing its role as a versatile biotechnological platform.

One key area of progress will be the discovery and optimization of regulatory elements. Just as other microbial hosts are benefitting from the availability of diverse promoters and finely tuned regulatory parts, *A. gossypii* will gain from the development of new inducible and repressible systems that allow precise control of gene expression. This is particularly important for genes whose overexpression negatively affects growth, where dynamic regulation can help balance productivity and fitness.

A similar parallel can be drawn in the area of protein secretion. In many fungi and yeasts, understanding the nuances of native *versus* heterologous signal peptides has been crucial for improving secretion efficiency. Extending this knowledge to *A. gossypii* will be essential for optimizing its use as a host for extracellular proteins and metabolites.

Genome editing is another area where trends across biotechnology point the way forward. CRISPR systems are becoming more efficient, multiplexed, and flexible, and the same trajectory is expected for *A. gossypii*. Beyond improving editing efficiency across different genomic loci, the implementation of CRISPR interference (CRISPRi) and activation (CRISPRa) will allow transcriptional modulation without the need for permanent genetic alterations. Cleaner editing strategies, leaving fewer scars, will also align with broader efforts to reduce genome burden and preserve strain robustness for industrial processes.

Just as importantly, the progress of *A. gossypii* will be shaped by community-wide efforts to share knowledge, standardize tools, and build open platforms, an approach that has already accelerated innovation in other microbial systems. Collaborative tool development will ensure that advances are not only achieved faster but also disseminated more broadly, enabling a growing number of researchers to contribute to and benefit from the expanding potential of this organism.

Taken together, these advances reflect both the general trajectory of biotechnology and the specific promise of *A. gossypii*. As molecular tools become increasingly precise and adaptable, this filamentous fungus is poised to emerge not only as a robust microbial cell factory but also as a model for how non-conventional organisms can be harnessed in the service of a sustainable and circular bioeconomy.

Recent works have already demonstrated *A. gossypii*'s ability to produce several novel metabolites of biotechnological interest along with riboflavin, such as lactones (Silva et al. 2019), microbial oils (Francisco et al. 2023), terpenes (Muñoz-Fernández et al. 2022), nucleosides (Ledesma-Amaro et al. 2015), orotic acid (Silva et al. 2022), aromatic alcohols (Ravasio et al. 2014), among others. While these results are promising, most metabolites still require further development to reach titers, yields, and productivities suitable for industrial application, whether through metabolic engineering or bioprocess optimization.

It is increasingly recognized that the choice of host chassis is a critical determinant of success in any metabolic engineering strategy (Cunha et al. 2015; Romaní et al. 2015), and that integrating this selection early in the bioprocess design is essential for achieving optimal outcomes (Costa et al. 2017). In this context, the biodiversity of *A. gossypii* strains isolated from diverse environments has begun to be explored (Dietrich et al. 2013; Li et al. 2025). To date, characterization has focused primarily on their genome sequences and structural features, including mating-type loci, chromosomal translocations, and gene duplications (Dietrich et al. 2013). Moving forward, a deeper understanding of the endogenous metabolism of novel isolates will be crucial. This would enable to identify the most suitable host chassis for a given target product, enabling more rational, efficient, and predictable engineering strategies that can accelerate the development of industrially robust *A. gossypii* platforms.

Furthermore, the establishment of such new products demands for a deeper understanding of the techno-economic performance of each application, which will ultimately dictate the viability and feasibility of their implementation. Along with this, the environmental impact of the production processes has also to be assessed, which is key to supporting viability when competing with chemical synthesis alternatives. These two aspects are of particular importance in the fostering of a bio-based economy, where low-cost substrates, such as agricultural and industrial by-products and wastes pose as promising low-cost and renewable feedstocks for the design of economically and environmentally sustainable biotechnological processes.

There are several promising emerging *A. gossypii* biosynthetic routes using industrial waste as feedstocks, e.g., crude glycerol (Silva et al. 2022), lignocellulosic hydrolysates (Francisco et al. 2023; Muñoz-Fernández et al. 2024), or oil wastes (Martín-González et al. 2025), which are still at an early stage of development. Thus, bioprocess engineering efforts focusing on the optimization of such waste-based processes are expected to continue, to foster the consolidation of economic, sustainable bioproduction with *A. gossypii*, ultimately translated in a robust platform with a wide coverage of substrates and products.

Adding to this, the pool of compounds produced in *A. gossypii* is expected to increase beyond the already known and described molecules, driven by the higher efficiency of novel molecular tools for *A. gossypii* manipulation and by the acknowledged robustness of this fungus as a biological chassis and its tunable metabolic traits. Moreover, rather than focusing on a single product, future processes are expected to increasingly pursue the simultaneous production of multiple metabolites, as exemplified by the approach proposed by Birk and colleagues for the concurrent synthesis of lactones and riboflavin (Birk et al. 2019). Such co-production strategies offer a promising route to improve the techno-economic balance of metabolites that are otherwise challenging to produce, by maximizing the overall value generated from a single cultivation. By enabling the valorization of multiple waste streams, these approaches not only enhance the feasibility of bioproduction but also contribute to more sustainable and resource-efficient processes, aligning with the broader goals of the circular bioeconomy.

Complementarily to the efforts being made upstream for the biosynthesis of added-value products, the translation of new molecules into novel applications will also demand the development of downstream processes for their recovery. In this sense, there is a significant space to be explored in what regards to the integration of the upstream and downstream sections of the bioprocess. The ability to tune *A. gossypii* toward targeted metabolic outputs will not only enhance titers and productivity but also define configuration, efficiency and cost of the downstream process. Also, there is still a considerable opportunity to be pursued in the development and implementation of extractive fermentations, which can have a direct impact not only on the cost and complexity of product recovery, but also in overall titers and yields achieved via biosynthesis. This approach has recently been explored for the production of γ-decalactone by bioconversion of ricinoleic acid using other organisms (Wang et al. 2024) and represents a promising approach to be explored for the production of several metabolites from *A. gossypii*.

Ultimately, the introduction of new *A. gossypii* based products will require an expansion of the regulatory directives and demonstration of their safety in food, feed and pharma applications. It will leverage on the previously acknowledged approval status of riboflavin, already approved by several regulatory organizations for animal feed and human applications, with very well-established requirements and guidelines (Rychen et al. 2018).

In conclusion, the future of *A. gossypii* biotechnology lies at the intersection of molecular innovation, strain diversity, and integrated bioprocess design. Continued advances in genetic and metabolic engineering tools, combined with a deeper understanding of the natural metabolism of diverse strains, will enable the rational selection and optimization of host chassis for specific target products. The expansion of multi-product and co-valorization strategies, together with the development of efficient upstream and downstream processes, will enhance both the economic and environmental sustainability of production. Equally important is the translation of these advances from laboratory to industrial scale: scaling up cultivation while maintaining productivity, robustness, and product quality will be essential for the successful implementation of *A. gossypii*-based processes. Finally, these scientific and technological developments must be aligned with regulatory frameworks to ensure the safe introduction of new products into food, feed, and pharmaceutical applications. Taken together, these perspectives highlight *A. gossypii* not only as a versatile and robust microbial platform but also as a cornerstone for the development of innovative, scalable, circular, and sustainable bio-based production systems in the coming decades.

References

Birk F, Fraatz MA, Esch P, Heiles S, Pelzer R, Zorn H (2019) Industrial riboflavin fermentation broths represent a diverse source of natural saturated and unsaturated lactones. J Agric Food Chem 67:13460–13469. https://doi.org/10.1021/acs.jafc.9b01154

Costa CE, Romaní A, Cunha JT, Johansson B, Domingues L (2017) Integrated approach for selecting efficient *Saccharomyces cerevisiae* for industrial lignocellulosic fermentations: importance of yeast chassis linked to process conditions. Bioresour Technol 227:24–34. https://doi.org/10.1016/j.biortech.2016.12.016

Cunha JT, Aguiar TQ, Romaní A, Oliveira C, Domingues L (2015) Contribution of *PRS3*, *RPB4* and *ZWF1* to the resistance of industrial *Saccharomyces cerevisiae* CCUG53310 and PE-2 strains to lignocellulosic hydrolysate-derived inhibitors. Bioresour Technol 191:7–16. https://doi.org/10.1016/j.biortech.2015.05.006

Dietrich FS, Voegelli S, Kuo S, Philippsen P (2013) Genomes of *Ashbya* fungi isolated from insects reveal four mating-type loci, numerous translocations, lack of transposons, and distinct gene duplications. G3 (Bethesda) 3:1225–1239. https://doi.org/10.1534/g3.112.002881

Francisco M, Aguiar TQ, Abreu G, Marques S, Gírio F, Domingues L (2023) Single-cell oil production by engineered *Ashbya gossypii* from non-detoxified lignocellulosic biomass hydrolysate. Fermentation 9:791. https://doi.org/10.3390/FERMENTATION9090791/S1

Ledesma-Amaro R, Buey RM, Revuelta JL (2015) Increased production of inosine and guanosine by means of metabolic engineering of the purine pathway in *Ashbya gossypii*. Microb Cell Fact 14:58. https://doi.org/10.1186/s12934-015-0234-4

Li CH, Ou JH, Chen CY (2025) *Eremothecium* species from *Cardiospermum halicacabum* and *Koelreuteria henryi*, and their associated soapberry bugs. Mycol Progress 24:55. https://doi.org/10.1007/s11557-025-02073-4

Martín-González J, Montero-Bullón JF, Muñoz-Fernández G, Buey RM, Jiménez A (2025) Valorization of waste cooking oil for bioproduction of industrially-relevant metabolites in *Ashbya gossypii*. N Biotechnol 88:32–38. https://doi.org/10.1016/J.NBT.2025.04.002

Muñoz-Fernández G, Martínez-Buey R, Revuelta JL, Jiménez A (2022) Metabolic engineering of *Ashbya gossypii* for limonene production from xylose. Biotechnol Biofuels Bioprod 15:79. https://doi.org/10.1186/s13068-022-02176-0

Muñoz-Fernández G, Montero-Bullón JF, Martínez JL, Buey RM, Jiménez A (2024) *Ashbya gossypii* as a versatile platform to produce sabinene from agro-industrial wastes. Fungal Biol Biotechnol 11:16. https://doi.org/10.1186/s40694-024-00186-1

Ravasio D, Wendland J, Walther A (2014) Major contribution of the Ehrlich pathway for 2-phenylethanol/rose flavor production in *Ashbya gossypii*. FEMS Yeast Res 14:833–844. https://doi.org/10.1111/1567-1364.12172

Romaní A, Pereira F, Johansson B, Domingues L (2015) Metabolic engineering of *Saccharomyces cerevisiae* ethanol strains PE-2 and CAT-1 for efficient lignocellulosic fermentation. Bioresour Technol 179:150–158. https://doi.org/10.1016/j.biortech.2014.12.020

Rychen G, Aquilina G, Azimonti G et al (2018) Safety and efficacy of vitamin B2 (riboflavin) produced by *Ashbya gossypii* DSM 23096 for all animal species based on a dossier submitted by BASF SE. EFSA J 16:e05337. https://doi.org/10.2903/J.EFSA.2018.5337

Silva R, Aguiar TQ, Coelho E, Jiménez A, Revuelta JL, Domingues L (2019) Metabolic engineering of *Ashbya gossypii* for deciphering the *de novo* biosynthesis of γ-lactones. Microb Cell Fact 18:62. https://doi.org/10.1186/s12934-019-1113-1

Silva R, Aguiar TQ, Domingues L (2022) Orotic acid production from crude glycerol by engineered *Ashbya gossypii*. Bioresour Technol Rep 17:100992. https://doi.org/10.1016/j.biteb.2022.100992

Wang W, Li W, Xin X, Liang J, Liu G, Bai W, Zhang M (2024) Bioconversion of ricinoleic acid to (R)-γ-decalactone by *Clavispora lusitaniae* YJ26 cells in an ionic liquid-containing biphasic fermentation system. Food Biosci 61:104953. https://doi.org/10.1016/J.FBIO.2024.104953